Advanced Technologies in Agricultural Engineering and Energy Optimization

Advanced Technologies in Agricultural Engineering and Energy Optimization

Vadim Bolshev
Vladimir Panchenko
Nallapaneni Manoj Kumar
Pandian Vasant
Igor Litvinchev
Prasun Chakrabarti

MDPI

Basel • Beijing • Wuhan • Barcelona • Belgrade • Novi Sad • Cluj • Manchester

Vadim Bolshev
Laboratory of Power Supply and Heat Supply
Federal Scientific Agroengineering Center VIM
Moscow
Russia

Vladimir Panchenko
Laboratory of Systems of Non-Traditional Energy
Federal Scientific Agroengineering Center VIM
Moscow
Russia

Nallapaneni Manoj Kumar
School of Energy and Environment
City University of Hong Kong
Kowloon
Hong Kong

Pandian Vasant
Faculty of Electrical and Electronic Engineering
Ton Duc Thang University
Ho Chi Minh City
Vietnam

Igor Litvinchev
Graduate Program in Systems Engineering
Nuevo Leon State University
Nuevo Leon
Mexico

Prasun Chakrabarti
Deputy Provost
ITM (SLS) Baroda University
Vadodara
India

Editorial Office
MDPI AG
St. Alban-Anlage 66
4052 Basel, Switzerland

This is a reprint of the Special Issue, published open access by the journal *Energies* (ISSN 1996-1073), freely accessible at: www.mdpi.com/journal/energies/special_issues/959Z967992.

For citation purposes, cite each article independently as indicated on the article page online and as indicated below:

Lastname, A.A.; Lastname, B.B. Article Title. *Journal Name* **Year**, *Volume Number*, Page Range.

ISBN 978-3-7258-0920-2 (Hbk)
ISBN 978-3-7258-0919-6 (PDF)
https://doi.org/10.3390/books978-3-7258-0919-6

Contents

About the Editors

Vadim Bolshev

Vadim Bolshev is a senior scientific officer with the Laboratory of Power and Heat Supply of the Federal Scientific Agroengineering Center in VIM, Russia. Vadim Bolshev received the M.S. degree in engineering from Orel State Agrarian University in 2012 and the Candidate of Technical Science degree from the Federal Scientific Agroengineering Center VIM in 2020. From 2010 to 2018, he gained experience in the industry as an electrician, an electrical engineer, and the chief engineer. Since 2018, he has been working as a researcher with the Laboratory of Power and Heat Supply, Federal Scientific Agroengineering Center VIM. His field of scientific activity is to develop methods and tools aimed at improving power supply efficiency, including the development of methods and devices for monitoring power quality and the technical state of power supply system elements.

Vladimir Panchenko

Vladimir Panchenko works as a professor at the Department of Theoretical and Applied Mechanics, Russian University of Transport, Moscow, Russia. He also holds a position at the Laboratory of Systems of Non-Traditional Energy, Federal Scientific Agroengineering Center VIM, Moscow, Russia.

Prof. Vladimir's research disciplines include renewable energy, agroengineering technologies, agricultural engineering, autonomous power supply, industrial design, electric transport, three-dimensional modeling, environmental engineering, and manufacturing engineering.

Nallapaneni Manoj Kumar

Dr. Nallapaneni is a transdisciplinary energy and sustainability engineer with a Ph.D. in Digital Circular Economy and Circular Power System from the School of Energy and Environment, City University of Hong Kong. He has obtained two Masters degrees, one in Renewable Energy Technologies from Karunya University, India, and the other in Environmental Economics from Annamalai University, India. He holds a bachelor's degree in Electrical and Electronics Engineering from GITAM University. Before joining CityU, he worked as a research fellow at Universiti Malaysia Pahang, Malaysia, on a project that focused on using solar photovoltaics as urban and rural infrastructure. Earlier, he worked as an assistant professor in the Department of Electrical and Electronics Engineering at the Bharat Institute of Engineering and Technology, Hyderabad, India, and as an energy engineer at Atiode Solar Systems Limited, Benin City, Nigeria.

Dr. Nallapaneni's research works focus on the topics of simulation, experimental, real-time empirical, and location- or climate-specific studies, mainly focused on building sustainable and resilient systems (decentralized, networked, and centralized) across critical infrastructure sectors by adopting nexus thinking and systems innovation with an emphasis on circular business models and digitalization. He worked on key sustainability challenges that include design, performance modeling, and analysis of a wide range of clean energy and environmental systems, the food–energy–water–waste nexus, industrial symbiosis, waste valorization and material passports, carbon accounting, and pricing. He possesses an interdisciplinary skill set that includes performance analytics, techno-economics, life cycle assessment, resilience assessment, leveraging digital innovation (blockchain, IoT, smart contracts, and AI), business model innovation, and nexus systems design with better conceptualization skills.

Pandian Vasant

Dr. Pandian Vasant is a research associate at the MERLIN Research Centre, Faculty of Electrical and Electronic Engineering, Ton Duc Thang University (TDTU), Vietnam. He holds a Ph.D. in Computational Intelligence from the Universidad Empresarial de Costa Rica, also known as UNEM, Costa Rica. He did his MSc in Engineering Mathematics from the University Malaysia Sabah, Malaysia, and his B.Sc. Hons. in Mathematics from the University of Malaya, Malaysia.

He served as the general chair of the International Conference on Intelligent Computing and Optimization (ICO). His research disciplines include artificial intelligence computing in mathematics, natural science, and engineering and medicine algorithms.

Igor Litvinchev

Igor Litvinchev received his M.Sc. degree in Applied Mathematics from the Moscow Institute of Physics and Technology (Fizteh), his Ph.D. in Systems Theory and Operations Research, and his Dr. Sci. (Habilitation) in Systems Modeling and Optimization from the Computing Center, Russian Academy of Sciences, Moscow. He is currently a research professor at Nuevo Leon State University (UANL), Mexico. His research is focused on large-scale systems modeling, optimization, and control with interdisciplinary applications. Professor Litvinchev is the author of 4 books and the editor of 12 more books published by Kluwer, Springer, and Elsevier. He published more than 90 research papers in leading international journals and served on the steering, program, and organizing committees for more than 70 international conferences. He also served as a guest editor for 8 journal special issues and was a reviewer for more than 60 prestigious journals. His research was supported by more than 30 grants from the NATO Scientific Affairs Division and European Community; ISF (USA) and RFBR (Russia); CNPq and FAPESP (Brasil); BRFBR (Belarus); and CONACYT, PROMEP, and PAICYT (Mexico). Professor Litvinchev is a member of the Russian Academy of Natural Sciences and the Mexican Academy of Sciences. He is a co-founder of the Mexican Society of Operations Research and the Mexican Logistics and Supply Chain Association.

Prasun Chakrabarti

Prof. Dr. Prasun Chakrabarti is a Deputy Provost at ITM SLS Baroda University, Vadodara, India. He received his Ph.D. (Eng.) from Jadavpur University in 2009; he also received his D.Litt. thesis from Sambalpur University in 2022. He was conferred with an Hon. D.Sc. from Shiraz University of Medical Sciences, Iran; an Hon. D.Eng. from the Iranian Neuroscience Society; and an Hon. Ph.D. from Lincoln University College Malaysia.

He supervised 11 Ph.D.s successfully. He authored and co-authored 100+ SCI/Scopus/Web of Science publications, 55 international granted patents, 1 granted Indian patent, 13 granted Indian copyrights, and 11 books. He is a Fellow of IEI, IETE, INSS, IET UK, and the Royal Society of Arts London. Moreover, he is a Senior Fellow of the Institute of Digital Information, University of South Wales, UK.

Preface

Major growth in population, along with its flow from rural areas to urban agglomerations, causes the problem of food shortages and becomes the primary determinant of global poverty. The food shortage is also significantly exacerbated by land degradation, which lowers agricultural yields and livestock productivity. Fortunately, a rapid development of technologies, including information technologies and renewable energy sources, is currently observed in many aspects of the economy. The direction of research is towards the utilization of renewable energy, which will help fulfill the energy demand and also mitigate environmental problems. The use of renewable energy plays an important role in agriculture, where technologies are also being improved from year to year; agricultural production is growing, and machinery and systems are becoming more autonomous and robotic, which is no longer possible without complex computing, optimization, planning, and working with large amounts of data. Nowadays, a large amount of unstructured heterogeneous data fuels the demand to extract useful insights in an automatic, reliable, and scalable way. The agriculture sector, however, is historically less receptive to innovation and lags behind the implementation of contemporary solutions, which defines the relevance of this Special Issue. The purpose of this reprint, titled "Advanced Technologies in Agricultural Engineering and Energy Optimization", is to publish research papers as well as review articles addressing recent advances in agriculture engineering within the confines of the use of various energy types. This reprint covers high-quality papers from academics and industry-related researchers in the areas of power supply in rural areas, biofuels and renewable energies used in agriculture, energy efficiency and conservation in agriculture, agricultural robotic applications, livestock production, application of electrophysical impact on agricultural objects, technologies in harvesting and seed machinery, solutions for digital and precision agriculture, applied mathematics, environmental bioengineering, machine learning, artificial intelligence, pattern recognition, data mining, multimedia processing, and big data to show the most recently advanced methods.

Vadim Bolshev, Vladimir Panchenko, Nallapaneni Manoj Kumar, Pandian Vasant, Igor Litvinchev, and Prasun Chakrabarti
Editors

energies

MDPI

Article

Landfill Waste Segregation Using Transfer and Ensemble Machine Learning: A Convolutional Neural Network Approach

Angelika Sita Ouedraogo, Ajay Kumar * and Ning Wang

Biosystems and Agricultural Engineering, Oklahoma State University, 111 Agriculture Hall, Stillwater, OK 74078, USA
* Correspondence: ajay.kumar@okstate.edu; Tel.: +1-405-744-5431

Abstract: Waste disposal remains a challenge due to land availability, and environmental and health issues related to the main disposal method, landfilling. Combining computer vision (machine learning) and robotics to sort waste is a cost-effective solution for landfilling activities limitation. The objective of this study was to combine transfer and ensemble learning to process collected waste images and classify landfill waste into nine classes. Pretrained CNN models (Inception–ResNet-v2, EfficientNetb3, and DenseNet201) were used as base models to develop the ensemble network, and three other single CNN models (Models 1, 2, and 3). The single network performances were compared to the ensemble model. The waste dataset, initially grouped in two classes, was obtained from Kaggle, and reorganized into nine classes. Classes with a low number of data were improved by downloading additional images from Google search. The Ensemble Model showed the highest prediction precision (90%) compared to the precision of Models 1, 2, and 3, 86%, 87%, and 88%, respectively. All models had difficulties predicting overlapping classes, such as glass and plastics, and wood and paper/cardboard. The environmental costs for the Ensemble network, and Models 2 and 3, approximately 15 g CO_2 equivalent per training, were lower than the 19.23 g CO_2 equivalent per training for Model 1.

Keywords: CNN; deep learning; ensemble learning; Inception–ResNet; EfficienNet; DenseNet; MSW; image classification; computational cost

Citation: Ouedraogo, A.S.; Kumar, A.; Wang, N. Landfill Waste Segregation Using Transfer and Ensemble Machine Learning: A Convolutional Neural Network Approach. *Energies* **2023**, *16*, 5980. https://doi.org/10.3390/en16165980

Academic Editors: Pandian Vasant, Vladimir Panchenko, Vadim Bolshev, Marcin Kaminski, Igor S. Litvinchev, Prasun Chakrabarti and Nallapaneni Manoj Kumar

Received: 5 July 2023
Revised: 3 August 2023
Accepted: 13 August 2023
Published: 15 August 2023

1. Introduction

Waste generation has increased over time along with demography, and soared from 88.1 million tons in 1960 to 292.4 million tons in 2018 [1]. The world waste production is expected to increase by 70% and reach 3.4 billion tons by 2050 [2]. Landfilling is the most common waste disposal method used worldwide, especially in the US. However, the constant increase in waste generation has raised concerns about land availability, and health and safety of humans, animals, and the environment [3]. The US Environmental Protection Agency (U.S EPA) provided the following waste management hierarchy: source reduction and reuse, recycling/composting, energy recovery, treatment, and disposal [1].

Waste recycling is one of the most environmentally friendly waste management processes proposed. The process allows material recovery and saves energy by avoiding new raw material mining (metals) and production (plastics, glass, and papers) [4]. In addition, recycling reduces landfilling activities and therefore minimizes air and water pollution [5]. However, waste recycling remains a challenge because of the lack of viable and cost-effective technologies for the segregation of wet and contaminated waste (waste with high moisture and food waste) [6]. Waste recycling involves manual and mechanical sorting. Manual sorting is inefficient and dangerous for workers because of the toxic nature of waste [6,7]. In addition, the lack of manpower makes the process challenging [6]. Mechanical sorting on the other hand processes dry recyclables only and requires presorting by households or at initial disposal place (trash can). However, the lack of public knowledge and awareness on waste recycling practices and improper waste categorization by residents often lead to inefficient recycling and result in contaminated batches that get routed to landfills [7,8]. Due to the aforementioned reasons, the recycling efficiency in the US is low and about half of the US waste is still landfilled. Hence, it is imperative to develop cost effective technologies that will not rely on household waste separation and be able to sort waste at landfill or processing sites.

The convolutional neural network (CNN) method, a subbranch of deep learning specifically for computer vision, has potential use in building a cost-effective waste segregation system for landfill wastes using classification of waste images. The availability of performant CNN algorithms helps towards developing cost effective and automated waste sorting systems with visually guided robotic arms for waste segregation. According to the US Environmental Protection Agency [1], landfill wastes are municipal solid wastes composed of over nine types of materials: paper and paperboard, glass, metals, plastics, yard trimmings, food, wood, rubber and leather, textiles, misc. inorganic waste, and others. These waste classes can be separated and reused (recyclable material) or properly disposed of via biological (composting and anaerobic digestion of food waste) and thermal processing (gasification and pyrolysis with energy recovery) to avoid landfilling. The main goal of this study is to develop and fine tune a CNN algorithm to classify landfill waste into nine marketable material classes (aluminum, carton, e-waste, glass, organic waste, paper and cardboard, plastics, textiles, and wood).

Several conference papers investigated waste classification using CNNs but very limited number of peer-reviewed published journals on the same topic exist. A pretrained CNN (ResNet 18) was used to classify recyclable waste into four classes (plastic, paper, metals, and glass) in one study that reported an accuracy of 87.8% [4]. Another study reported an accuracy of 82% in classifying recyclable waste into five classes using another pretrained model, DenseNet169 [7]. These studies only targeted recyclables and categorized waste into four and five classes. In this study, the landfill waste will be classified into nine classes and a combination of CNNs, transfer learning, and ensemble learning will be used. Moreover, based on the results reported in literature, the accuracy of waste classification using the CNN can be improved [4,7]. To date, no study has used a combination of Inception–ResNet, EfficientNet, and DenseNet via transfer and ensemble learning to classify landfill waste into nine classes. Consequently, the three pretrained CNNs were used as a base model to develop the ensemble network. This study investigated and compared the proposed model's performance to the three other networks (Models 1, 2, and 3) trained on the same data and conditions as the ensemble model. In addition, the models' training cost and environmental burden were investigated.

The rest of the paper is presented in four sections: Section 2 defines the CNN base models used; Section 3 presents the methodology used; Section 4 presents the results and discussion, and Section 5 presents the summary and conclusions.

2. Pretrained CNN Models Architecture

This study used three pretrained CNNs as base models to build the ensemble network. These pretrained Convolutional Neural Networks are models that were trained on large datasets. For example, ImageNet was created by a team of researchers to provide a large database for object recognition models training [9]. The database is composed of more than a million images categorized into 1000 classes and has been used in the ImageNet Large Scale Visual Recognition Challenge (ILSVRC) every year since 2010. The challenge has allowed the development of several CNN models such as VGG (Visual Geometry Group) and Inception (previously called GoogleNet) [10]. Pretrained CNN models have shown a potential for image classification. In this study, the pretrained models including Inception–ResNet-v2, EfficientNetB3, and DenseNet201 were used to develop an ensemble model for landfill waste classification into nine classes.

2.1. Inception–ResNet

The Inception–ResNet-v2, introduced by Szegedy, et al. [11], is a hybrid version based on the architectures of Inception family and residual connection [11]. The author described the network as costly but significantly performant in terms of recognition [11]. Inception–ResNet-v2 has a deep structure composed of 164 layers. As shown in Figure 1, the model's basic architecture is composed of stem block, five Inception–ResNet-As, a reduction-A, ten Inception–ResNet-Bs, a reduction-B, five Inception–ResNet-Cs, an average pooling, and a dropout layer (Figure 1) [11]. The residual connection used in the model is known for improving the model training speed and reducing network degradation due to its deep structure [11,12].

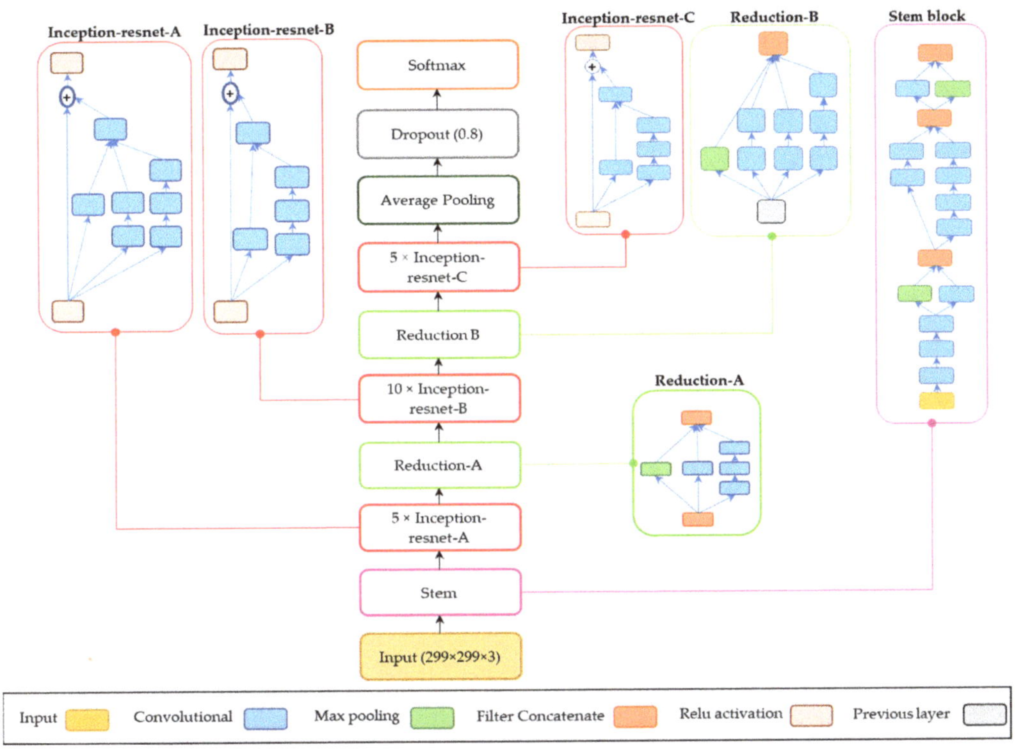

Figure 1. Inception–ResNet-v2 architecture inspired from [11].

2.2. EfficientNet

EfficientNet was introduced by Tan and Le [13]. The network was reported to be 8.4 times smaller and 6.1 times faster than existing CNNs. Tan and Le [13] applied advanced scaling to all three dimensions (depth, width, and resolution) of the network using an effective compound coefficient. Unlike other research that scale up depth, width, or resolution, the authors uniformly scaled up the three dimensions with a fixed ratio, which therefore, led to a higher accuracy (Figure 2) [13,14].

2.3. Densely Connected Convolutional Network (DenseNet)

The Densely Connected Convolutional Network (DenseNet) comes from a collaborative work between Cornell and Tsinghua Universities and Facebook AI research [15]. The network was designed to address challenges related to gradient vanishing due to the increased depth of the CNN models [15]. The authors found that connecting all layers to each other guarantees information flow throughout the network. Figure 3 shows the architecture of DenseNet 201 (201 layers), which is composed of a convolution layer, pooling layer, dense blocks, and transition layers (convolution and average pooling layers).

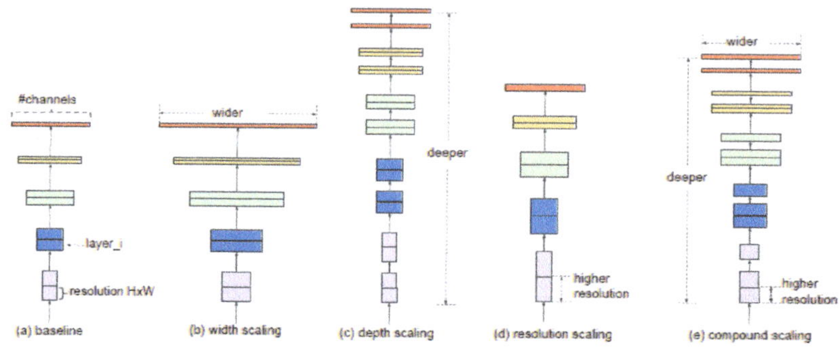

Figure 2. Model Scaling. (**a**) a baseline network example; (**b–d**) conventional scaling that only increases one dimension of network width, depth, or resolution. (**e**) proposed compound scaling method that uniformly scales all three dimensions with a fixed ratio [13].

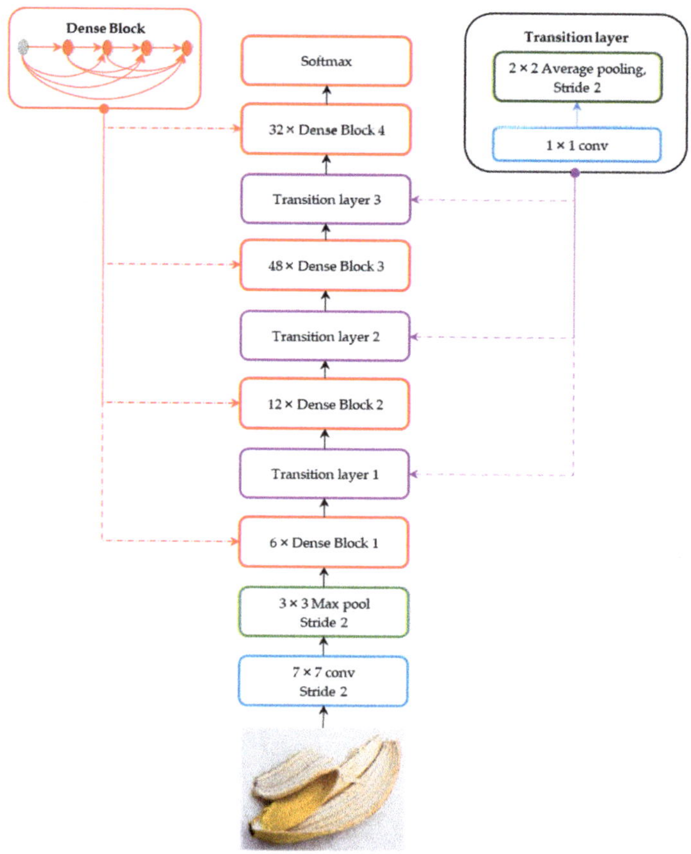

Figure 3. DenseNet 201 architecture.

3. Methods

This study proposed a CNN model using transfer and ensemble learning to classify landfill waste into nine classes: aluminum, carton, e-waste, glass, organic waste, paper and cardboard, plastics, textiles, and wood.

3.1. Deep Learning Libraries

Open-source platforms and libraries such as Keras, TensorFlow, Numpy, Matplotlib, Scikit-Learn, and Seaborn were used to train the models, and Google Colab Pro [16] was used as the training platform. Keras is a deep learning framework and open-source library capable of running on top of TensorFlow [17]. Keras provides full access to the TensorFlow platform and can run experiments faster [17]. Tensorflow is an open-source library able to carry complex numerical computation for machine learning and artificial intelligence [17]. In this study, Tensorflow and Keras were used for data processing (ImageDataGenerator), CNN architecture (layers, model), model training/optimization (regularizers, optimizers and callbacks), and evaluation (utils). Numpy is a library for Python used to work with arrays, matrices, linear algebra, and Fourier transform, etc. In this study, Numpy was used to transform images into matrices. Matplotlib is a library and an extension of Numpy used for plotting. The library was used to plot models, training curves, and performance metrics. The Scikit-Learn library was used to establish the models' classification reports (performance metrics) and confusion matrices. Google Colab Pro was used as a training platform for the CNN models. The platform offers faster GPU (NVIDIA P100 or T4), longer runtimes, and additional RAM.

3.2. Data Collection and Preprocessing

A waste dataset of 22,500 images was collected from waste classification dataset on Kaggle [18]. The dataset was preprocessed through image scrapping (repeated and misclassified image removal) and reorganized into nine groups with a total of 6536 images: aluminum, carton, e-waste, glass, organic waste, paper and cardboard, plastics, textiles and wood. Waste classes such as e-waste, carton, textiles, and wood had a low number of images. Therefore, the aforementioned classes datasets were increased by downloading 1810 images from Google Search and a recycling waste dataset on Kaggle. Finally, 8346 images were included in the study. Figure 4 and Table 1 show the dataset samples and repartition per class of waste, respectively.

Figure 4. Sample of images from the collected dataset.

Table 1. Waste dataset collection and repartition per class.

№	Classes	Class Items	Initial Database	Added Database	Total Number of Images
1	Aluminum	Canes, plates, bottles, leads, bottle openers, trash cans, cooking pots, car parts, and silverware	1019	-	1019
2	Carton	Juice, milk, and cigarettes boxes	382	151	533
3	E-waste	Batteries, electronics (computer, phones, etc.) circuit boards, microchips, cables, and chargers	-	1029	1029
4	Glass	Bottles, jars, containers, cups, decoration, plates, and pitchers	1089	-	1089
5	Organic waste	Fruits, vegetables, meats, fast food, meals, plants, seeds, cheese, bread, and eggshells	1053	-	1053
6	Paper and cardboard	Newspapers, magazines, books, shipping boxes, letters, envelopes, gift and pizza boxes, shredded paper, flyers, and stickers.	1194	-	1194
7	Plastics	Bottles, containers, cups, plates, food packaging, bags, silverware, furniture, cases, buckets, planting pots, and trash bins	1035	-	1035
8	Textiles	Clothes, curtains, towels, decorations, sheets, bags and fabric	346	484	830
9	Wood	Signs, furniture, cases, wood blocks, tiles, utensils, plates, silverware, wine cork, pellets, boards, baskets, mashed wood, and containers.	418	146	564
	Total		6536	1810	8346

After the dataset was uploaded in the simulation platform, all images in RGB format were resized to 224 × 224 pixels in resolution for data uniformity. A data augmentation technique is used to increase dataset size, reduce overfitting, capture more features, and therefore, increase the CNN models' performance. In this study, the ImageDataGenerator function was used for data augmentation techniques such as horizontal flip, shearing (0.2), zooming (0.2), and dataset repartition (training and validation). The waste dataset was then divided into training dataset (80%) and validation dataset (20%).

3.3. *Ensemble Method*

The method proposed in this study is called the ensemble method. The technique consists of combining feature extraction techniques of three CNN models to improve waste class prediction using the concatenation function. Transfer learning is known as a suitable technique to address the lack of data and computing cost. The hypothesis of this study is that combining several CNN models using transfer learning and ensemble learning will enhance useful and diverse feature collection and increase waste prediction accuracy, while reducing misclassification errors (for classes with similar features such as plastics, glasses, and metals). The pretrained models were trained on the waste dataset and the optimized weights were used to build the ensemble model.

The ensemble model architecture can be divided into three sections (Figure 5). The first section (grey in Figure 5) is the image collection and preprocessing step. The second section (Model 1, 2, and 3) consists of using weights of three pretrained CNNs through transfer learning to classify the waste dataset into nine classes. Models 1, 2, and 3 were built using Inception–ResNet, EfficientNetb3, and DenseNet201 as base models, respectively. An input layer was created for each model to define the image shape. The models' feature extraction abilities were optimized using layers such as a batch normalization layer, dense layer, dropout layer, and an additional dense layer (classifier for nine classes). Once the models were trained on the waste dataset, the updated weights were ensembled through concatenation (third section) and trained again (yellow in Figure 5). The networks' loss function, also called its objective function, was defined using categorical crossentropy (multiple classes). The optimizer in charge of the networks' learning rate was set using Adamax (0.001 initial learning rate). When CNN models are trained independently, the concatenation step is challenging and leads to errors because layers have repetitive names and parameters are not compatible. To resolve the issue, a method was created to allow Models 1, 2, and 3 to be trained and validated in the same algorithm. The models were saved and used to build, train, and validate the ensemble model.

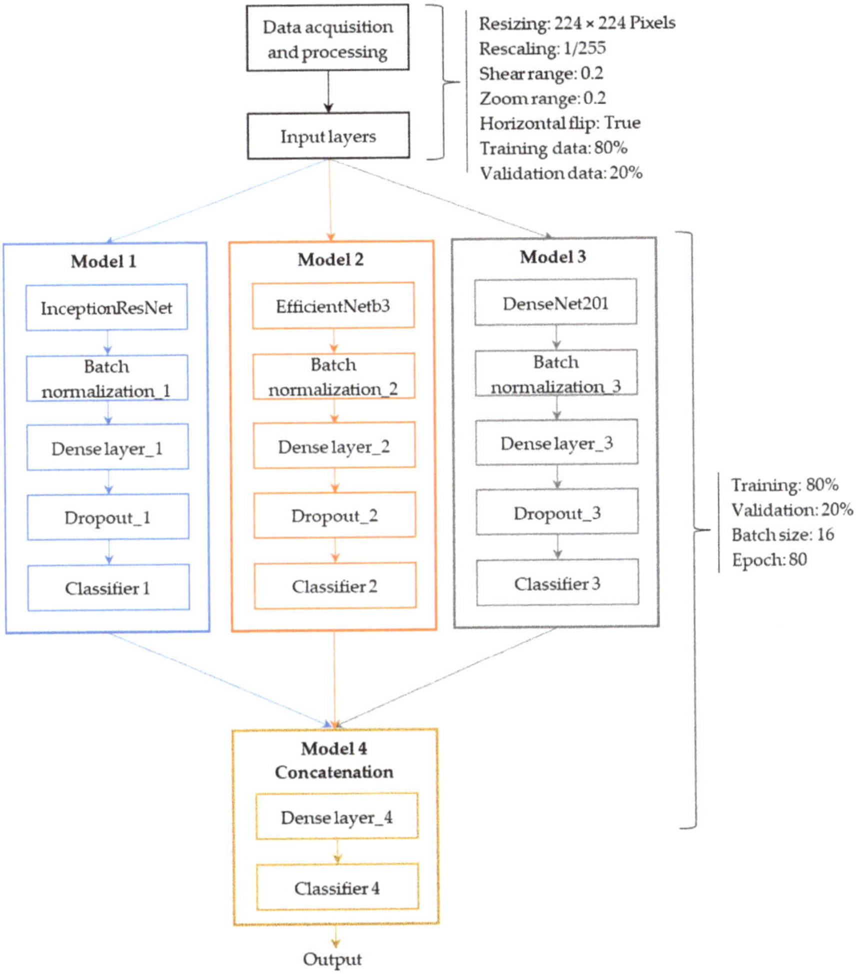

Figure 5. Ensemble model architecture.

3.4. Experiment Setting

Four CNN models (Model 1: Inception–ResNet-v2 based, Model 2: EfficientNetB3 based, Model 3: DenseNet201 based, and Model 4: the Ensemble Model) were run with the waste dataset of 8346 images containing nine classes of waste. CNNs. The training was completed on 80% of the waste dataset and the remaining 20% was used for testing and validation. The networks were trained in 80 epochs.

3.5. Performance Measures

The performance of each CNN model was evaluated using accuracy, precision, f-1 score, and recall. Performance metrics were evaluated using prediction indicators, true and false positives and true and false negatives. A true positive is when a data point belonging to a positive class is correctly predicted (belongs to positive class). A true negative is when a data point belonging to a negative class is correctly predicted (belongs to negative class). Alternatively, a false positive and negative correspond to an incorrect prediction of a positive and negative class, respectively.

Accuracy is the number of correct predictions made over the total predictions (Equation (1)),

$$Accuracy\ (\%) = \frac{True\ positives + True\ negatives}{Total\ prediction}. \tag{1}$$

Precision is the ratio of true positives and total positives (Equation (2)),

$$Precision\ (\%) = \frac{True\ Positives}{(True\ positives\ + False\ positives)}. \tag{2}$$

Recall or sensitivity is the quotient of true positive and sum of true positive and false negatives (Equation (3)),

$$Recall\ (\%) = \frac{True\ Positives}{(True\ positives\ + False\ negatives)}. \tag{3}$$

F1-score is the harmonic mean of precision and recall (Equation (4)),

$$F - 1\ score\ (\%) = \frac{2 \times Precision \times Recall}{Precision + Recall}. \tag{4}$$

4. Results and Discussion

4.1. Performance Metrics

Figure 6 shows the accuracy, precision, recall, and F1-score of the Ensemble Model and the three single networks (Models 1, 2, and 3). The Ensemble Model was the most performant model (accuracy: 90% and precision: 90%) and was followed by Model 3 (accuracy: 88% and precision: 88%). Model 2 (accuracy: 87% and precision: 87%) and Model 1 (accuracy: 86% and precision: 86%) were the poorest performing models. As shown in Table 2, the Ensemble Model predicted each waste class did better than Models 1, 2, or 3. The precision accuracy for wood was very low for all models. However, the Ensemble Model prediction accuracy was higher (71% precision) than those of Models 1, 2, and 3 (69%, 63%, and 70%, respectively). The low precision accuracy for wood is due to the small data size of the class (Table 1). Another reason for the misclassification of wood as food or cardboard was due to feature similarities among the classes. The prediction accuracy of the model proposed in this study was higher than the results reported by Gyawali, Regmi, Shakya, Gautam and Shrestha [4] (88% accuracy) and Zhang, Yang, Zhang, Bao, Su and Liu [7] (82% accuracy). These results proved that combining multiple pretrained CNNs as base model increased feature extractions abilities and led to higher prediction accuracy. The effect of waste class number on the Ensemble Model's performance was investigated by training and testing the model to predict six waste classes. The model showed a prediction accuracy of 93%, leading to the conclusion that the model's performance increases as the number of classes decreases.

4.2. Error Per Class and Model

Wood, textiles, paper/cardboard, and plastics were the waste classes with the highest prediction errors. The prediction errors were calculated by summing misclassified images per class and per model. As mentioned above, all models performed poorly in classifying wood, with precision values of 69, 63, 70, and 71% for Models 1, 2, and 3 and the Ensemble

Model, respectively. Prediction performances for Models 1 and 2 were low for plastics (82 and 83%, respectively), while Model 3 showed a low precision for paper and cardboard class (86%). Figure 7 shows the percentage of errors for each model per class. Overall, the Ensemble Model had the lowest prediction error. However, this model was the second-most accurate in predicting glass (behind Model 3), textiles (behind Model 1) and wood (behind Model 3).

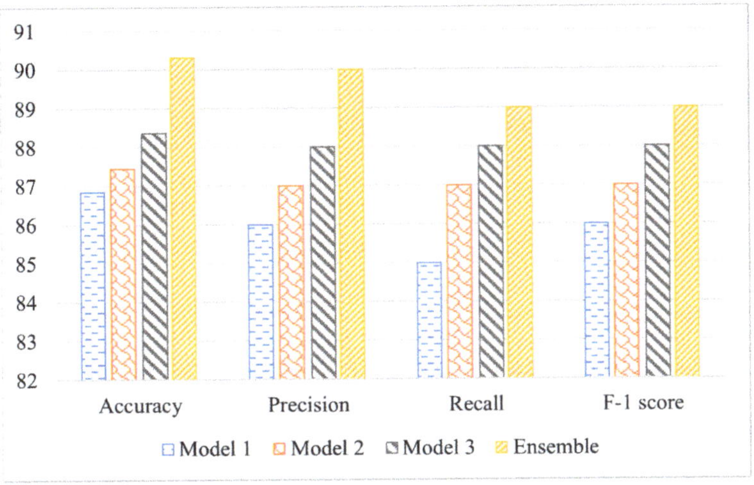

Figure 6. Performance metrics on validation data per model (%).

Table 2. Models' precision per class (%).

	Ensemble	Model 1	Model 2	Model 3
Aluminum	95	88	94	91
Carton	95	90	91	88
E-waste	93	92	90	92
Glass	93	92	91	92
Organic waste	92	88	90	89
Paper & cardboard	88	87	85	86
Plastics	87	82	83	89
Textiles	95	86	95	92
Wood	71	69	63	70

4.3. Confusion Matrix

The confusion matrices (Figure 8) show the models' prediction performance on the test dataset. The horizontal axis represents the predicted values (predicted classes) from the CNN models and the vertical axis shows the true values (true classes) of the data. The diagonal line represents accurate predictions. Although the Ensemble Model's overall performance was higher than those of Models 1, 2, and 3, the results showed that all models had difficulties in the classification of waste classes with similar features such as glass and plastics, paper/cardboard and wood, e-waste and aluminum and wood and organic waste. Azis, et al. [19] reported that plastics were confused with glass and cardboard. Susanth, et al. [20] confirmed that glass was misclassified as metal and plastic, metal as glass, plastic as glass, and metal and paper as trash. Rahman, et al. [21], Mao, et al. [22], and this study observed similar trends. According to Huang, Lei, Jiao and Zhong [6], the misclassification errors could be due to several issues such as: 1. plastic and glass bottle were so similar that the human eye could not detect a difference; 2. a plastic or glass bottle was covered with a plastic or paper label; and 3. metal was covered with a plastic or paper sticker. Wood and textiles were additional classes sorted in this study. Wood was mostly misclassified as paper and carboard because of feature similarities. Mao, Chen, Wang and Lin [22] supported that paper and carboard features were extracted based on the edges and corners. Wood shares similar features.

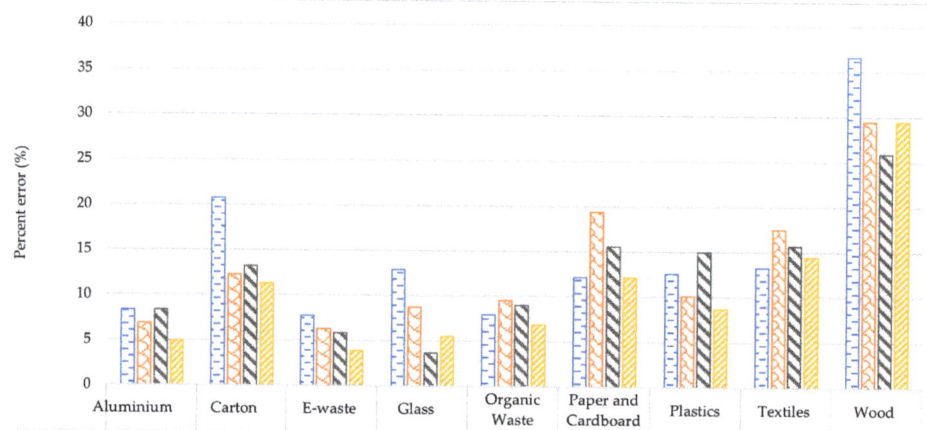

	Aluminium	Carton	E-waste	Glass	Organic Waste	Paper and Cardboard	Plastics	Textiles	Wood
Model 1	8.37	20.75	7.80	12.90	7.98	12.18	12.56	13.25	36.61
Model 2	6.90	12.26	6.34	8.76	9.57	19.33	10.14	17.47	29.46
Model 3	8.37	13.21	5.85	3.69	9.04	15.55	14.98	15.66	25.89
Ensemble	4.93	11.32	3.90	5.53	6.91	12.18	8.70	14.46	29.46

Figure 7. Classification error on validation set per class and model.

This study was a unique case study using three CNN models' knowledge to classify landfill waste into nine classes. Limited number of peer reviewed journals in the field have reported such a comparison. Table 3 shows the comparison between other research findings (4 conferences and 2 journals), and the results obtained with the ensemble method. The Ensemble method's accuracy was higher than the accuracies reported by Miko, et al. [23] (75%) with Inception v3 and Ruiz, et al. [24] with ResNet (89%). Accuracy reported by other studies varied between 93 and 95% when models were trained to classify six classes (Table 3), while this study classified nine waste classes. An evaluation of the Ensemble Model on six waste classes showed a higher accuracy (93%) than on nine waste classes, which was among the highest. Wood misclassification led to lower nine-class prediction accuracy. In addition, the characteristics of the dataset (non-uniform background, different color light, and non-obvious features) used to train and test the models affected the results. Overall, the Ensemble Model prediction performance was higher than the pretrained CNNs investigated in this study. The results of that model proved that the combination of transfer and ensemble learning reduced the sensitivity of CNNs to small datasets and increased useful feature extraction.

Trained CNN models are dependent on image datasets that the model was trained on, as the model's learning process is based on features extracted from the images. Waste image prediction is challenging because of image noises such as background, object deformation, dirt, and presence of several types of waste on an image. To increase the prediction accuracy of the models presented in this study, images with several backgrounds and deformed images were included in the dataset. Though these types of noise were included in the dataset used in this study, other noise types, such as wastes with soil/dirt, were not included.

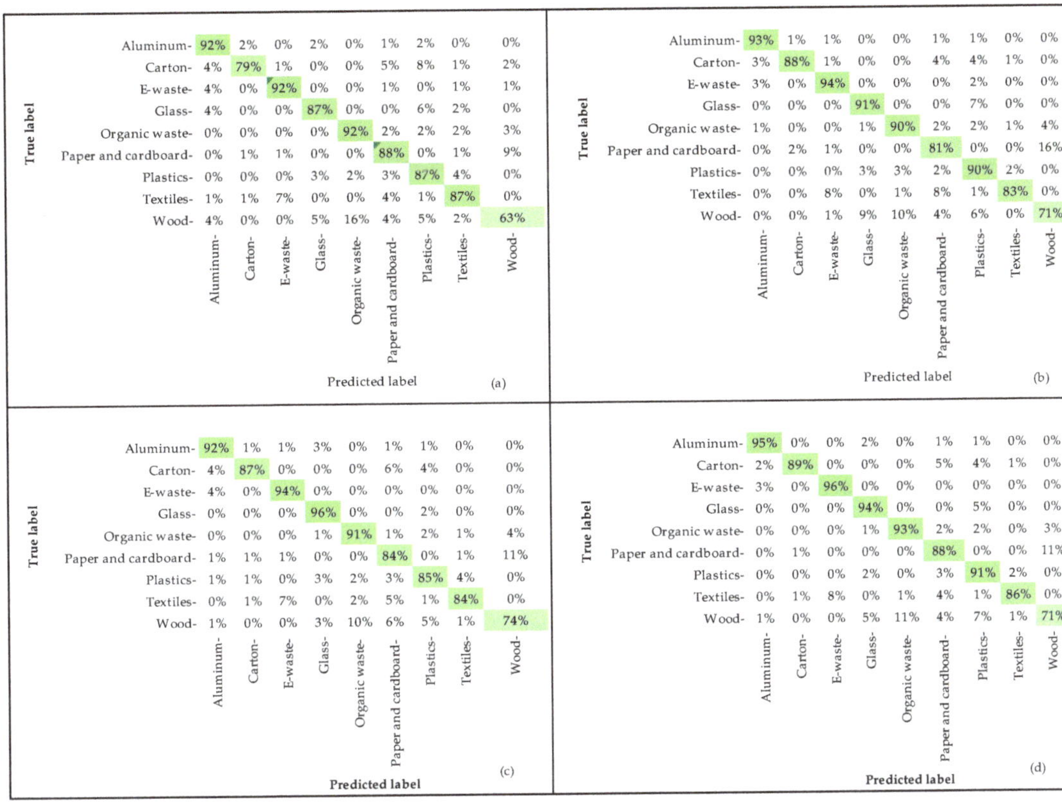

Figure 8. Confusion matrices: (**a**) Model 1; (**b**) Model 2; (**c**) Model 3; (**d**) Ensemble.

Table 3. Comparative results of different models.

#	Method	Data Source	Data Size	Number of Classes	Classes	Accuracy (%)	References
1	Inception V3	-	2433	6	Cardboard, glass, paper, plastic, metal, and organic waste	75	[23]
2	ResNet	TrashNet	2527	6	Cardboard, glass, paper, plastic, metal, and trash	89	[24]
3	Inception-v3	GitHub	2400	6	Cardboard, glass, paper, plastic, metal, and others	93	[19]
4	YOLO	-	2527	6	Cardboard, glass, paper, plastic, metal, and organic trash	94	[25]
5	DenseNet169	TrashNet and Google images	4163	6	Cardboard, glass, paper, plastic, metal, and trash	95	[20]
6	ResNet-34	GITHUB	2560	6	Cardboard, glass, paper, plastic, metal, and trash	95	[21]
7	Ensemble	Kaggle and Google images	5559	6	Cardboard, glass, paper, plastic, aluminum, and organic waste	93	This study
8	Ensemble	Kaggle and Google images	8346	9	Paper and cardboard, glass, plastic, aluminum, organic waste, carton, wood, textiles, and e-waste	90	This study

4.4. Models Training Cost

Training CNNs requires large computational resources with fast GPUs. Although cost can vary with time and between vendors, the average value of a Tesla T4 GPU is $1797 [26]. This study trained and tested the ensemble model on a paid subscription of Google Colab Pro using GPU (Tesla T4). Use of computational resources to train CNNs requires energy consumption and therefore leads to an environmental cost. Using the method proposed in Strubell, et al. [27], the carbon footprint of training the CNN models was calculated. The total power consumption (Pt) of each model shown in (Equation (5)) was calculated using each model's training time (t), the 2022 power usage effectiveness factor (1.55) [28], and the power (Pg) consumed by the Tesla T4 GPU (70 W) [29]. The environmental cost is the amount of CO_2 emitted due to training the CNN models and was calculated using the factor provided by the US EPA for average CO_2 produced per power consumed (0.976 lb. of CO_2 equivalent/kWh) [30]. Table 4 shows the training time, total power consumption and environmental cost (grams CO_2 equivalent) of the Ensemble network and Models 1, 2, and 3.

$$P_t = 1.58\, tP_g \tag{5}$$

Table 4. CNNs models computational costs.

CNN Models	Training Time (Minutes)	Total Power (Wh)	Environmental Cost (g CO_2 Equivalent)
Ensemble	19.28	34.87	15.45
Model 1	24.00	43.40	19.23
Model 2	19.57	35.38	15.68
Model 3	18.77	33.94	15.04

According to Table 4, training the Ensemble networks emitted less carbon dioxide (15.45 g CO_2 equivalent) compared to Models 1 (19.23 g CO_2 equivalent) and 2 (15.68 g CO_2 equivalent). However, Model 3 showed the lowest carbon footprint (15.04 g CO_2 equivalent). The Ensemble Model was built using three networks. However, the Ensemble Model's training time, power consumption, and environmental cost are close to, or lower than the single networks' computational cost. These results proved that the combination of transfer and ensemble learning was energy-efficient.

5. Conclusions

The exponential increase in waste generation, shortage of land availability, and environmental and health related issues have led to the search for novel waste management methods to limit landfilling. Waste recycling is one of the most preferred methods for waste management. However, due to lack of cost-effective sorting technologies, waste segregation and recycling remain a challenge. Advances in machine learning have the potential to solve this challenge through development of automated and visually guided robotic arms to sort wastes. In this study, several models (Models 1, 2, and 3) were developed with pretrained Inception-ResNet-v2, EfficientNetB3, and DenseNet201 as base models using transfer learning. An ensemble model was developed using a combination of the three models via transfer and ensemble learning. The performance metrics showed the ensemble model was the highest performant of all networks, with a precision of 90%, while precision ranged from 86% to 88% for Models 1, 2, and 3. The results showed that by combining transfer and ensemble learning approaches, network performance increased and increased essential feature extraction despite, the relatively small dataset. Additionally, the multi-network's environmental cost (15.45 g CO_2 equivalent) was similar to single networks' (Models 2 and 3) cost (15.68 and 15.04 g CO_2 equivalent, respectively) and lower than Model 1′s environmental impact (19.23 g CO_2 equivalent).

Author Contributions: Conceptualization, A.S.O., A.K. and N.W.; methodology, A.S.O.; software, A.S.O.; validation, A.K. and N.W.; formal analysis, A.S.O.; investigation, A.S.O.; resources, A.K. and N.W.; data curation, A.S.O.; writing—original draft preparation, A.S.O.; writing—review and editing, A.K. and N.W.; visualization, A.S.O., A.K. and N.W.; supervision, A.K. and N.W.; project administration, A.K.; funding acquisition, A.K. All authors have read and agreed to the published version of the manuscript.

Funding: This research was funded in part by the OSU Research Foundation, Oklahoma Agricultural Experiment Station, and the USDA National Institute of Food and Agriculture, Hatch project 1019511.

Data Availability Statement: Publicly available datasets were analyzed in this study. This data can be found here: https://www.kaggle.com/datasets/angelikasita/waste-images.

Conflicts of Interest: The authors declare no conflict of interest.

References

1. US Environmental Protection Agency. National Overview: Facts and Figures on Materials, Wastes and Recycling. Available online: https://www.epa.gov/facts-and-figures-about-materials-waste-and-recycling/national-overview-facts-and-figures-materials (accessed on 1 March 2021).
2. The World Bank. Solid Waste Management. Available online: https://www.worldbank.org/en/topic/urbandevelopment/brief/solid-waste-management (accessed on 24 March 2022).
3. Ouedraogo, A.S.; Frazier, R.S.; Kumar, A. Comparative Life Cycle Assessment of Gasification and Landfilling for Disposal of Municipal Solid Wastes. *Energies* **2021**, *14*, 7032. [CrossRef]
4. Gyawali, D.; Regmi, A.; Shakya, A.; Gautam, A.; Shrestha, S. Comparative analysis of multiple deep CNN models for waste classification. *arXiv* **2020**, arXiv:2004.02168.
5. Chu, Y.; Huang, C.; Xie, X.; Tan, B.; Kamal, S.; Xiong, X. Multilayer hybrid deep-learning method for waste classification and recycling. *Comput. Intell. Neurosci.* **2018**, *2018*, 5060857. [CrossRef] [PubMed]
6. Huang, K.; Lei, H.; Jiao, Z.; Zhong, Z. Recycling waste classification using vision transformer on portable device. *Sustainability* **2021**, *13*, 11572. [CrossRef]
7. Zhang, Q.; Yang, Q.; Zhang, X.; Bao, Q.; Su, J.; Liu, X. Waste image classification based on transfer learning and convolutional neural network. *Waste Manag.* **2021**, *135*, 150–157. [CrossRef] [PubMed]
8. Deng, J.; Xu, W.-Y.; Zhou, C.-B. Investigation of waste classification and collection actual effect and the study of long acting management in the community of Beijing. *Huan Jing Ke Xue* **2013**, *34*, 395–400. [PubMed]
9. Li, F.-F.J.D.; Olga, R.; Alex, B.; Kai, L. About ImageNet. Available online: https://www.image-net.org/about.php (accessed on 14 September 2022).
10. Russakovsky, O.; Deng, J.; Su, H.; Krause, J.; Satheesh, S.; Ma, S.; Huang, Z.; Karpathy, A.; Khosla, A.; Bernstein, M. Imagenet large scale visual recognition challenge. *Int. J. Comput. Vis.* **2015**, *115*, 211–252. [CrossRef]
11. Szegedy, C.; Ioffe, S.; Vanhoucke, V.; Alemi, A.A. Inception-v4, inception-resnet and the impact of residual connections on learning. In Proceedings of the Thirty-First AAAI Conference on Artificial Intelligence, San Francisco, CA, USA, 4–9 February 2017.
12. Elhamraoui, Z. InceptionResNetV2 Simple Introduction. Available online: https://medium.com/@zahraelhamraoui1997/inceptionresnetv2-simple-introduction-9a2000edcdb6 (accessed on 7 July 2022).
13. Tan, M.; Le, Q. Efficientnet: Rethinking model scaling for convolutional neural networks. In Proceedings of the International Conference on Machine Learning, Long Beach, CA, USA, 9–15 June 2019; pp. 6105–6114.
14. Zhang, P.; Yang, L.; Li, D. EfficientNet-B4-Ranger: A novel method for greenhouse cucumber disease recognition under natural complex environment. *Comput. Electron. Agric.* **2020**, *176*, 105652. [CrossRef]
15. Huang, G.; Liu, Z.; Van Der Maaten, L.; Weinberger, K.Q. Densely connected convolutional networks. In Proceedings of the IEEE Conference on Computer Vision and Pattern Recognition, Honolulu, HI, USA, 21–26 July 2017; pp. 4700–4708.
16. Google. *Welcome to Colaboratory Google (Google Colab)*; Google: Mountain View, CA, USA, 2022.
17. Keras. About Keras. Available online: https://keras.io/about/ (accessed on 14 September 2022).
18. Sekar, S. *Waste Classification Date*; Kaggle: San Francisco, CA, USA, 2019.
19. Azis, F.A.; Suhaimi, H.; Abas, E. Waste classification using convolutional neural network. In Proceedings of the 2020 2nd International Conference on Information Technology and Computer Communications, Online, 12–14 August 2020; pp. 9–13.
20. Susanth, G.S.; Livingston, L.J.; Livingston, L.A. Garbage Waste Segregation Using Deep Learning Techniques. In Proceedings of the IOP Conference Series: Materials Science and Engineering, Chennai, India, 2–3 October 2020; p. 012040.
21. Rahman, M.W.; Islam, R.; Hasan, A.; Bithi, N.I.; Hasan, M.M.; Rahman, M.M. Intelligent waste management system using deep learning with IoT. *J. King Saud Univ.-Comput. Inf. Sci.* **2020**, *34*, 2072–2087. [CrossRef]
22. Mao, W.-L.; Chen, W.-C.; Wang, C.-T.; Lin, Y.-H. Recycling waste classification using optimized convolutional neural network. *Resour. Conserv. Recycl.* **2021**, *164*, 105132. [CrossRef]

23. Miko, P.; Shiela, M.M.; Hilary, R.; Mary, J.S. i-BIN: An Intelligent Trash Bin for Automatic Waste Segregation and Monitoring System. In Proceedings of the 2019 IEEE 11th International Conference on Humanoid, Nanotechnology, Information Technology, Communication and Control, Environment, and Management (HNICEM), Laoag, Philippines, 29 November–1 December 2019.
24. Ruiz, V.; Sánchez, Á.; Vélez, J.F.; Raducanu, B. Automatic image-based waste classification. In Proceedings of the International Work-Conference on the Interplay Between Natural and Artificial Computation, Almería, Spain, 3–7 June 2019; pp. 422–431.
25. Geetha, S.; Saha, J.; Dasgupta, I.; Bera, R.; Lawal, I.A.; Kadry, S. Design of Waste Management System Using Ensemble Neural Networks. *Designs* **2022**, *6*, 27. [CrossRef]
26. David. Alternative to Colab Pro: Comparing Google's Jupiter Notebooks to Gradient Notebooks (Updated). Available online: https://blog.paperspace.com/alternative-to-google-colab-pro/ (accessed on 26 December 2022).
27. Strubell, E.; Ganesh, A.; McCallum, A. Energy and policy considerations for deep learning in NLP. *arXiv* **2019**, arXiv:1906.02243.
28. Davis, J.; Bizo, D.; Lawrence, A.; Rogers, O.; Smolaks, M. *Uptime Institute Lobal Data Center Urvey 2022*; Uptime Institute: New York, NY, USA, 2022.
29. Techpowerup. Tesla T4 Specs. Available online: https://www.techpowerup.com/gpu-specs/tesla-t4.c3316 (accessed on 26 December 2022).
30. U.S Environmental Protection Agency. Emissions & Generation Resource Integrated Database (eGRID). Available online: https://www.epa.gov/egrid (accessed on 26 December 2022).

energies

MDPI

Article

Deep Learning-Based Transformer Moisture Diagnostics Using Long Short-Term Memory Networks

Aniket Vatsa [1], Ananda Shankar Hati [1,*], Vadim Bolshev [2,*], Alexander Vinogradov [2], Vladimir Panchenko [3] and Prasun Chakrabarti [4]

1 Department of Electrical Engineering, Indian Institute of Technology (Indian School of Mines), Dhanbad 826004, India
2 Laboratory of Power Supply and Heat Supply, Federal Scientific Agroengineering Centre VIM, 109428 Moscow, Russia
3 Department of Theoretical and Applied Mechanics, Russian University of Transport, 127994 Moscow, Russia
4 Department of Computer Science and Engineering, ITM SLS Baroda University, Vadodara 391510, India
* Correspondence: anandashati@iitism.ac.in (A.S.H.); valdimbolshev@gmail.com (V.B.)

Abstract: Power transformers play a crucial role in maintaining the stability and reliability of energy systems. Accurate moisture assessment of transformer oil-paper insulation is critical for ensuring safe operating conditions and power transformers' longevity in large interconnected electrical grids. The moisture can be predicted and quantified by extracting moisture-sensitive dielectric feature parameters. This article suggests a deep learning technique for transformer moisture diagnostics based on long short-term memory (LSTM) networks. The proposed method was tested using a dataset of transformer oil moisture readings, and the analysis revealed that the LSTM network performed well in diagnosing oil insulation moisture. The method's performance was assessed using various metrics, such as R-squared, mean absolute error, mean squared error, root mean squared error, and mean signed difference. The performance of the proposed model was also compared with linear regression and random forest (RF) models to evaluate its effectiveness. It was determined that the proposed method outperformed traditional methods in terms of accuracy and efficiency. This investigation demonstrates the potential of a deep learning approach for identifying transformer oil insulation moisture with a R^2 value of 0.899, thus providing a valuable tool for power system operators to monitor and manage the integrity of their transformer fleet.

Keywords: power transformer; oil-immersed insulation; moisture forecasting; long short-term memory

Citation: Vatsa, A.; Hati, A.S.; Bolshev, V.; Vinogradov, A.; Panchenko, V.; Chakrabarti, P. Deep Learning-Based Transformer Moisture Diagnostics Using Long Short-Term Memory Networks. *Energies* **2023**, *16*, 2382. https://doi.org/10.3390/en16052382

Academic Editor: K. T. Chau

Received: 24 January 2023
Revised: 17 February 2023
Accepted: 28 February 2023
Published: 2 March 2023

1. Introduction

The prognostic health management (PHM) of transformers is essential since it enables the early detection of possible faults, thus preventing unanticipated downtime and expensive repairs [1]. PHM can also help optimize the transformers' maintenance schedule, which reduces costs and improves overall reliability. Additionally, PHM also helps to extend the lifespan of transformers by identifying and addressing potential issues at embryonic stages. Consequently, PHM contributes to transformers' safe and dependable operation, which is essential for preserving the power supply and preventing power outages.

The transformer's insulation can deteriorate due to moisture, which could result in arcing and electrical discharge [2]. It causes corrosion in the transformer tank, which lowers insulating performance, thus potentially leading the transformer to break down. When a significant moisture content is present, paper insulation ages more quickly, decreasing transformers insulating efficiency and life expectancy [3]. According to existing literature, moisture is not only responsible for the loss of the dielectric strength of the transformers, but it also upsurges power loss due to deterioration in insulating qualities, such as an increase in acidity and a fall in the flash point of transformer oil. The insulating performance of transformers is dynamic, which has been immensely quantified in the existing literature by estimating moisture and aging degrees [4]. Due to the detrimental effects moisture has on a transformer's operation and lifespan, assessing moisture levels is a crucial part of transformer condition monitoring processes.

Transformer oil-paper insulation is used in electrical transformers to insulate and protect electrical components from moisture. Moisture in the transformer oil-paper insulation causes a reduction in the transformer's efficiency, which leads to electrical failures. Thus, moisture forecasting in transformer oil-paper insulation helps to predict and prevent these issues and save energy. Maintenance schedules can be optimized by forecasting the moisture levels in transformer oil-paper insulation, preventing moisture from reaching harmful levels [5]. Therefore moisture forecasting can help prolong the transformer's life and reduce the prerequisite for repairs or replacements, ultimately saving energy [6]. Additionally, by monitoring and controlling the moisture levels in the transformer, the transformer's efficiency can be maintained, which in turn will help to reduce energy losses and improve the overall performance of the electrical grid. Furthermore, moisture forecasting can be used to anticipate the likelihood of moisture-induced defects, which can be used to plan transformer maintenance. Therefore, the danger of unplanned outages is reduced, and energy distribution can be planned appropriately.

Several methods for predicting moisture in transformer oil-paper insulation have been put forth in the literature. Dielectric frequency domain spectroscopy (DFDS), a non-destructive testing method used to assess the state of transformer oil-paper insulation, is one of the most frequently employed techniques [7]. The loss factor (tan delta), which is associated with the transformer oil-paper insulation's dielectric properties, is a measurement of the energy dissipation of the insulation. Another technique that has been proposed in the literature is the usage of sensors, such as capacitive sensors [8], resistive sensors [9], and optical sensors [10]. The moisture level in the oil-paper insulation of a transformer can be measured with these sensors. However, the intrusive nature of these sensors can compromise the transformer's oil-paper insulation integrity. Recently, machine learning-based approaches have been utilized for predictive health monitoring in transformer oil-paper insulation [11]. These methods predict the crucial factor in the transformer oil-paper insulation by using features obtained from the dielectric properties as inputs to a machine learning model, such as neural networks. Some studies have shown that these machine learning-based approaches can effectively anticipate crucial health-sensitive parameters and can be applied for the prognostic health monitoring of transformer oil-paper insulation. Implementing predictive maintenance is anticipating when moisture levels may be high and taking precautions before problems arise; this is done by analyzing historical information and applying machine learning algorithms. In terms of managing numerous correlated time series, long short-term memory (LSTM), a form of recurrent neural network (RNN), is frequently seen as being superior at prediction compared to conventional statistical techniques, such as autoregressive integrated moving average (ARIMA) [12,13]. Additionally, LSTM networks can learn and represent non-linear correlations in the data, which is challenging to perform with conventional statistical techniques. Based on the content of existing literature, a novel method of a moisture determination-based DFDS test conducted on a small number of samples is presented in this paper. The findings suggest that the reported technique can predict oil-paper insulation moisture to obtain feature parameters for training the LSTM model. The key novelty of this research is to adopt a hybrid DFDS-LSTM-based approach to transformer moisture diagnosis.

2. Materials and Methods

The current density within insulation can be expressed in terms of the conduction current, the vacuum, and polarization displacement currents. The transition from the time to the frequency domain can be achieved analytically by the Fourier transform. The relationship between the total current density, $J(t)$, and the electric field intensity, $E(t)$, within an insulation system is expressed in the s domain as follows [14]:

$$J(\omega) = \sigma_o \overline{E}(\omega) + j\omega\epsilon_o \overline{E}(\omega) + j\omega\epsilon_o \overline{F}(\omega)\overline{E}(\omega) \quad (1)$$

The Fourier transform of the dielectric response function is denoted as $\overline{F}(\omega)$, which is equivalent to the complex susceptibility $\overline{\chi}(\omega)$ [15], where $\overline{\chi}(\omega)$ can be represented as imaginary and real components ($\overline{\chi}(\omega) = \chi'(\omega) - j\,\chi''(\omega)$) as follows:

$$J(\omega) = \omega\epsilon_o \left\{ \left[\frac{\sigma_o}{\omega\epsilon_o} + \chi''(\omega) \right] + j[1 + \chi'(\omega)] \right\} \overline{E}(\omega) \quad (2)$$

In terms of the complex relative dielectric permeability, ($\bar{\varepsilon}_r = \epsilon_r'(\omega) - j\,\epsilon_r''(\omega)$) can be expressed as follows:

$$J(\omega) = j\omega\epsilon_o [\epsilon_r''(\omega) - j\,\epsilon_r''(\omega)] \overline{E}(\omega) \quad (3)$$

By comparing Equations (2) and (3), the frequency-dependent dielectric dissipation factor can be given as follows:

$$tan\delta = \frac{\epsilon_r''(\omega)}{\epsilon_r'(\omega)} = \frac{\frac{\sigma_o}{\omega\epsilon_o} + \chi''(\omega)}{1 + \chi'(\omega)} \quad (4)$$

Transformer oil-paper insulation is examined using DFDS, a non-destructive testing method. As shown in Figure 1, the procedure comprises supplying the transformer with a variable-frequency alternating current (AC) voltage and measuring the transformer oil-paper insulation's complex permittivity and loss factor in the frequency domain. The transformer oil-paper insulation has abnormalities, including moisture retention, partial discharge, and other impurities, which can be found and located using the DFDS procedure, which is a robust tool for monitoring the insulation condition.

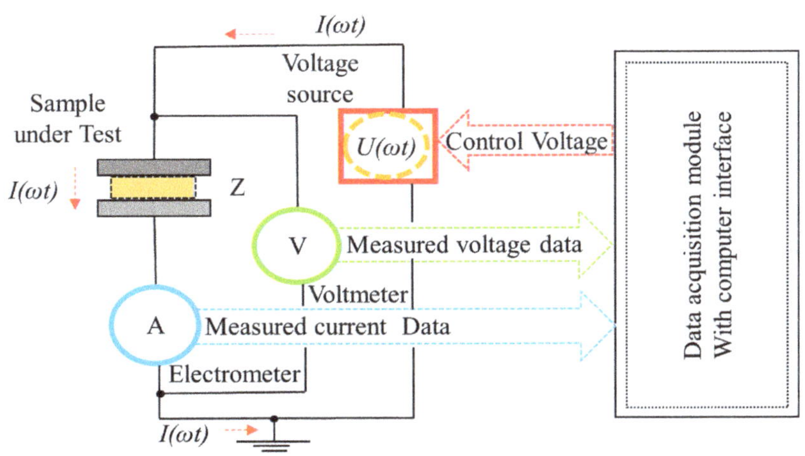

Figure 1. Schematic diagram of the dielectric frequency domain spectroscopy (DFDS) system.

The loss factor and the energy loss of the transformer insulation can also be directly correlated with energy dissipation.

The simplified expression for the energy loss per unit volume (E) can be expressed as:

$$E = 2\pi f * \varepsilon_0 \varepsilon_r'' * E_o^2 \quad (5)$$

where f is the frequency, and the electric field strength is represented by E_o.

2.1. Accelerated Aging Experiment

The insulation samples used for the accelerated aging of transformer oil-paper insulation samples are prepared in a laboratory by simulating the conditions that cause aging in real-world transformers. A flowchart of the experiment design for extracting the DFDS dataset for the transformer moisture diagnostics technique is shown in Figure 2. The experimental setup for conducting accelerated thermal aging and dielectric frequency domain spectroscopy (DFDS) analysis of various aging samples is demonstrated in Figure 3a. The OPI aging samples use copper foil to replicate the high and low-voltage transformer windings (C, D). The composite insulation system is also simulated using a pressboard cylinder (A) and kraft paper (B, E); pressboard strips are utilized to separate the B and E kraft and to imitate oil ducts, as presented in Figure 3b. The insulation samples were constructed using kraft paper, a pressboard cylinder, and mineral oil with a density of 0.89 g/cm^3 at 20 °C.

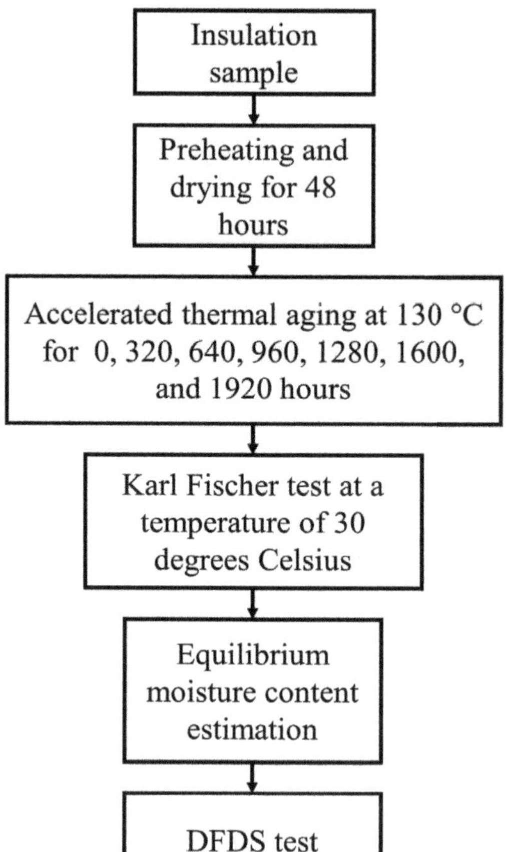

Figure 2. Schematic illustrating the experimental technique.

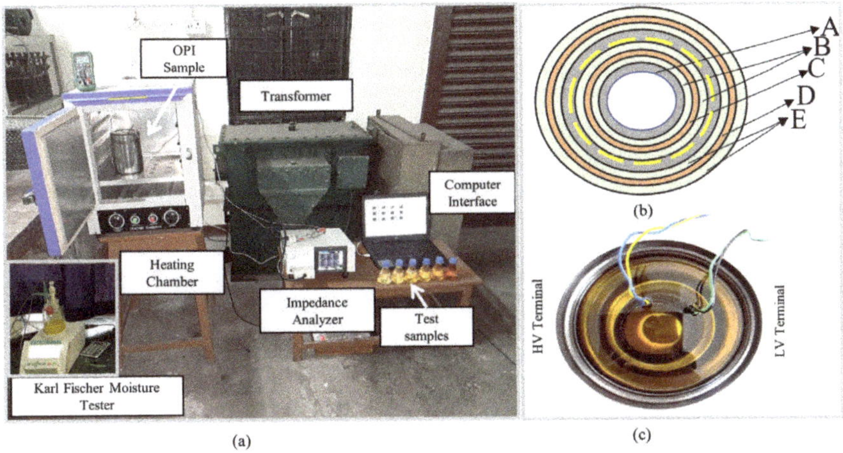

Figure 3. (**a**) Laboratory DFDS test setup with insulation sample and test oil samples, (**b**) Schematic view of the test sample, (c) Prepared sample.

In this study, 20 laboratory-prepared insulation samples were examined to analyze moisture variation within transformer oil-paper insulation. The samples underwent a preheating process for 48 h to remove the initial moisture content. The study utilized 14 of the 20 samples for training and 6 for validation and testing. Seven of the samples were subjected to accelerated aging at intervals of 0, 320, 640, 960, 1280, 1600, and 1920 h, and the same process was repeated for the remaining seven training samples. Notably, the training data was obtained from the 14 samples with an initial moisture content of less than 0.5%. To validate the results, accelerated aging samples were collected for testing and validation by taking out samples at various time intervals, up to 1920 h. In this research study, the oil moisture content was measured in parts per million (PPM) using the Karl Fischer test at a temperature of 30 degrees Celsius. Based on the results, the moisture content in the paper was estimated using Oommen moisture equilibrium curves. Subsequently, the dielectric frequency response of the insulation samples was measured using an impedance analyzer instrument to generate the DFDS dataset. The range of tan delta ($tan\delta$) measurements in the study was between 0.00225 and 0.383, which provides valuable insights into the electrical properties of the transformer oil-paper insulation.

The radar plot in Figure 4 provides valuable insights into the aging behavior of the tan delta values. It can be determined that the tan delta is higher in the low-frequency regions, except for the initial samples with low moisture content. Additionally, the plot reveals that the tan delta of the samples increases as the aging process advances in a clockwise direction. It is apparent that the aging conditions affect the dielectric properties of the insulation, specifically in the lower frequencies, and it helps monitor the state of the transformer oil-paper insulation and identify potential problems before they lead to significant failures. It is clearly evident from Figure 4 that the variation in tan delta with aging at varying frequencies is barely noticeable. This fact highlights the difficulty in extracting moisture-sensitive features from the dielectric frequency domain spectroscopy data using conventional statistical methods.

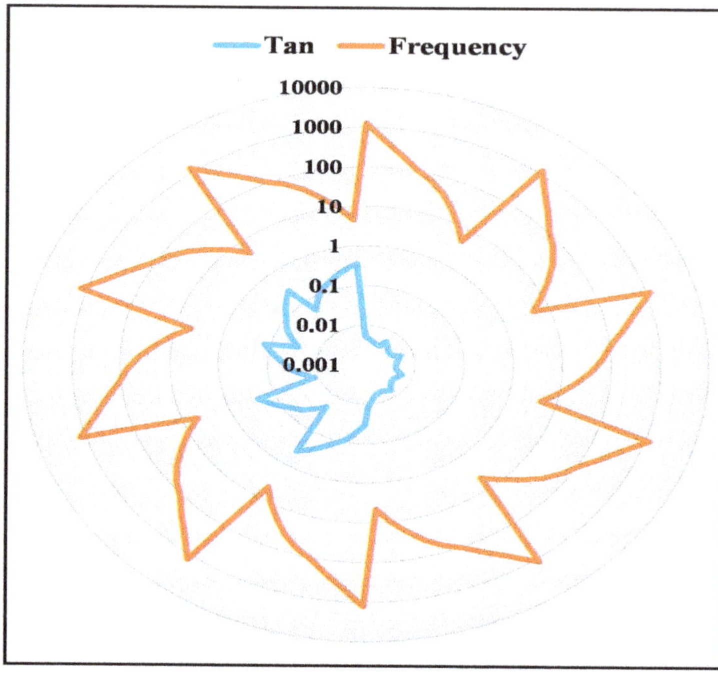

Figure 4. Radar plot of frequency and tan delta variations over the aging of transformer oil-paper insulation.

Moisture content is an essential indicator of the condition and aging of the transformer oil-paper insulation, and it is crucial to forecast it accurately. However, the stability of the dielectric properties makes it difficult to extract moisture-sensitive features from the DFDS data. Therefore, this paper proposes using an LSTM-based deep learning method for forecasting moisture variation inside the transformer tank through non-intrusive DFDS feature parameters. LSTM is a recurrent neural network that can capture temporal dependencies in the data, allowing it to make accurate predictions even when the data is not easily interpretable using conventional statistical methods [16]. Although insulation DFDS is also affected by the aging duration, the literature suggests that the effect of moisture on DFDS is significantly higher than aging [17]. This disparity becomes even smaller for oil-immersed insulation, thus justifying the proposed approach for moisture estimation through learning non-linear temporal dependencies using LSTM. Several accelerated aging tests of varying durations were carried out to study the impact of thermal aging on transformer oil. Utilizing a coulombmeter setup based on the Karl Fischer titration (KFT) technique, the moisture content of the oil is estimated. An impedance analyzer instrument was employed to measure the curves' DFDS characteristics of insulation samples. Predictive maintenance for transformer moisture assessment using machine learning algorithms involves several steps:

- Collecting historical data on moisture levels in the transformer oil and paper insulation while relevant variables.
- Data preprocessing through cleaning and transforming the data to remove any missing or inaccurate values.
- Training the ML model through preprocessed data to predict moisture levels based on the input variables.
- Validating the model through a separate dataset ensures generality and accurate prediction of different moisture levels.

2.2. Proposed Model

Long Short-term Memory (LSTM) is a recurrent neural network that excels at forecasting time series data. LSTMs have been demonstrated to perform effectively well on time series forecasting tasks. Furthermore, they have been employed in various industrial applications, including power system monitoring and control [18]. LSTMs are intended to respond with sequential data with many dependencies and long-term trends. An LSTM's central concept is to utilize gates to control the flow of information across the network, allowing it to recall specific information for more extended periods. An LSTM model is trained using specific information and data, which in this investigation includes historical data of oil moisture measurements, dielectric loss factor, aging hour, and cyclic features. These features are then extracted from the data to predict insulation conditions in terms of forecasting transformer oil moisture levels.

The input-output equation of an LSTM model can be represented as follows:

$$y(t) = f(x(t), h(t-1), c(t-1)) \tag{6}$$

where $y(t)$, $x(t)$, $h(t-1)$, $c(t-1)$, and $f(.)$ is the output at time step t, the input at time step t, the hidden state at time step t-1, the memory cell state at time step t-1, and the LSTM function, respectively. The LSTM function maps the input, hidden state, and memory cell state to the output at time step t [19]. The LSTM model takes in the input $x(t)$ at each time step t and uses it along with the hidden state $h(t-1)$ and memory cell state $c(t-1)$ from the previous time step to calculate the output $y(t)$ at the current time step. The hidden state and memory cell state is updated at each time step based on the input and previous hidden state, allowing the LSTM to maintain a certain level of information for an extended period. In more detail, the LSTM model takes the input features, the DFDS, and the aging hours (AH), and processes them through multiple layers of LSTM cells and fully connected layers. The LSTM cells capture the temporal dependencies in the data, while the fully connected layers map the input features to the output, which is the moisture concentration. Figure 5 provides a visual representation of the comprehensive training process for the whole moisture diagnosis algorithm. The training operation of an LSTM-based model for forecasting moisture variation inside the transformer tank using aging hour and tan delta as feature parameters involves several steps:

- Data preparation: the first step is to prepare the data for training, which includes collecting and preprocessing the data, such as cleaning, normalizing, and segmenting the data. The data is divided into input and output sets, with the aging hour and tan delta as the input features and moisture as the target or output.
- Model architecture: the LSTM model architecture is defined, including the number of layers, the number of units in each layer, and the activation functions used. The model architecture is designed to handle the input data, with the input layer reshaped to match the shape of the input data and the output layer designed to output the moisture values.
- Training: the model is then trained using the input-output data sets, where the input is the aging hour and tan delta data, and the output is the corresponding moisture content. The model is trained to learn the non-linear relationship between the input features and the output and make accurate moisture content predictions.
- Optimization: during the training process, the model is optimized using a chosen optimization algorithm, such as Adadelta, to minimize the error between the predicted and actual moisture content. The learning rate of the optimizer is a hyperparameter that is adjusted to optimize the model's performance.
- Evaluation: the model is evaluated using a validation set to assess its performance and accuracy after the training process, and it is fine-tuned by adjusting the hyperparameters.

- Forecasting: once the model is trained and optimized, it can forecast moisture content in transformer oil-paper insulation. The model takes an aging hour and tan delta as inputs and forecasts the moisture content.

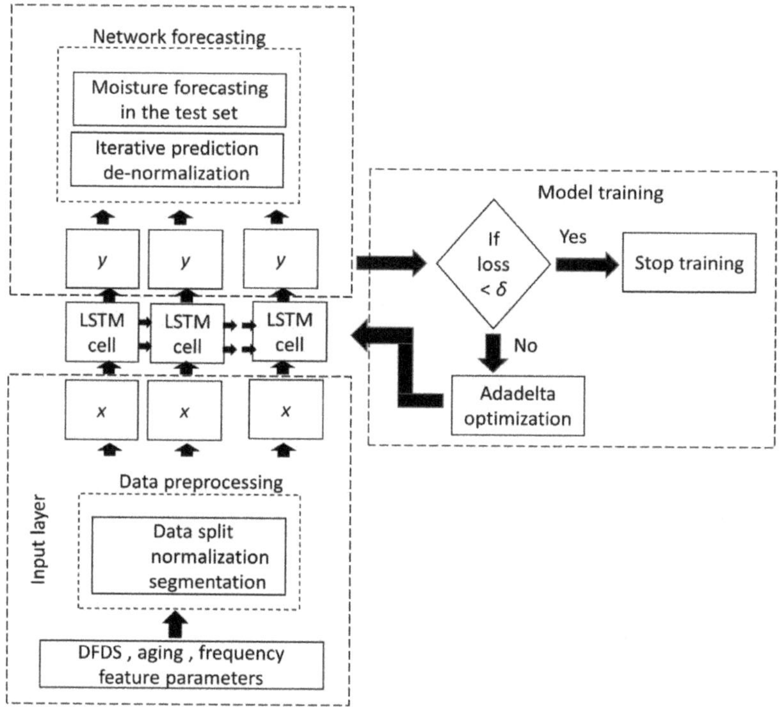

Figure 5. The flowchart of the moisture diagnosis algorithm.

The model architecture and hyperparameters, determined empirically in this study, are fine-tuned during the training phase to achieve the best outcomes. In this investigation, the moisture content of transformers was predicted using a long short-term memory network model. The model's architecture comprised an input layer, an LSTM layer, and a dense output layer. The input layer had the shape of (5, 6), which is implemented using the Keras library. This input layer reshaped the input data to a specific shape that the LSTM layer could handle. The LSTM layer had 100 units, and it is responsible for capturing the temporal dependencies in the data. The loss function used in the study was the mean absolute error (MAE).

A sigmoid activation function, a prevalent choice for this type of problem, was utilized by the LSTM layer. The sigmoid function is utilized to introduce non-linearity to the model, enabling it to learn intricate connections between the input and output properties. The output layer is a dense layer with a single unit that maps the output of the LSTM layer to the final predicted moisture content. The dense layer is connected to the LSTM layer with fully connected weights, allowing it to make a final prediction based on the output of the LSTM layer. The LSTM model was trained to utilize a series of input-output pairs consisting of dielectric frequency domain spectroscopy and aging hour data as inputs and the corresponding moisture content as the output. Through utilizing this technique, accurate moisture forecasting was acquired for oil-paper transformer insulation.

2.3. Model Optimization

Recurrent neural networks, such as LSTM, primarily excel at time series forecasting. However, adequate tuning of the LSTM model is required for excellent performance. Hyperparameters are adjustable parameters of a machine learning model that are established before the training process and do not undergo modification during training. A combination of hyperparameters, including the number of LSTM layers, learning rate, and sequence length, were employed in a deep-learning model to predict transformer insulation conditions through moisture forecasting. The selection of the optimal hyperparameters was achieved using a combination of techniques, such as grid search, error, and the reduce on plateau method.

The Adadelta optimizer is a popular method for tuning LSTM models. Adadelta is an adaptive learning rate optimization algorithm that modifies the learning rate automatically during training. It makes use of the gradient and past data to adjust the learning rate on the run. Adadelta is very effective for deep neural network training, such as LSTM, because it can help overcome diminishing or exploding gradients [20]. In this study, an Adadelta optimizer was utilized to tune the LSTM model. The initial learning rate was set to 0.01, and ρ and ε values were kept to 0.98 and 1×10^{-8}, respectively. The optimal learning rate that yields the best performance is chosen using a grid search method by evaluating its performance on a validation set. Also, Adadelta has the advantage of requiring less memory and processing than other optimization techniques, such as Adam, RMSprop, and Adagrad. Furthermore, it is less sensitive to the initial learning rate, making it more resistant to hyperparameter selection [21]. The Adadelta optimizer increased the performance of the proposed LSTM model for obtaining accurate moisture forecasting results in transformer oil-paper insulation.

3. Results

The proposed LSTM-based transformer's insulation moisture forecasting model was trained and evaluated on a compiled dataset of dielectric frequency domain spectroscopy measurements of oil-paper insulation, which empowers the model to predict future moisture levels in the insulation accurately. During the model training, 70% of the dataset was used, while the remaining 30% was split into a 15% test set and a 15% validation set, the former was used to evaluate the model's performance, and the latter was used for hyperparameter tuning and overfitting prevention. An LSTM model on the dielectric aging DFDS dataset is trained in this study, and its performance is assessed by depicting the loss curve. The loss curve shown in Figure 6 demonstrates that the model's error lowers consistently as the training iterations are increased, demonstrating that the system effectively learns from the training data. The final training loss achieved by the model is 0.027, which is relatively low compared to the range of possible loss values. Additionally, the curve also exhibits a progressive decline in fluctuations, indicating that the model is not overfitting the training set of data. These outcomes show how well the LSTM model generalizes well to new information and, therefore, is effective at learning the features in the dataset. After the training operation is complete, the model validation is carried out independently on a previously intact sample. Quantifiable metrics, including the R-squared (R^2) value, are used to assess the performance and generalization capabilities of the proposed moisture level forecasting model. R-squared is a statistical measure that shows the percentage of the variance in the dependent variable that can be predicted by the independent variable. R-squared is typically used to measure the fraction of the variance in observed oil moisture levels that are predictable from the historical data provided as input to the LSTM model when forecasting transformer oil moisture.

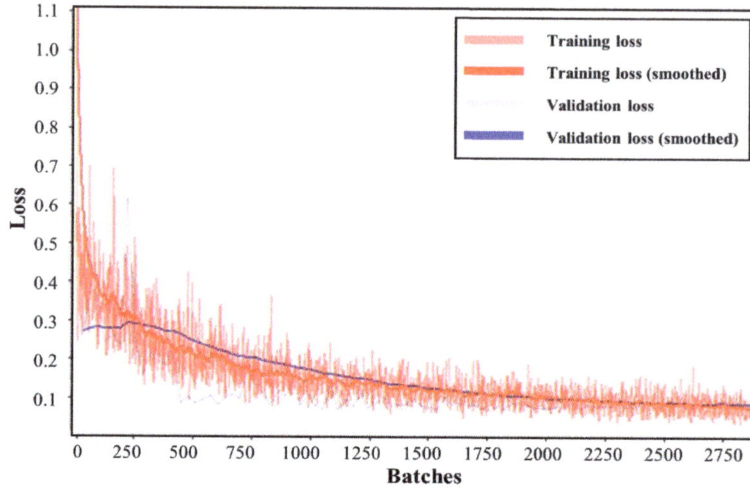

Figure 6. Loss curve vs. batches curve of the LSTM forecasting model.

The results of the LSTM-based moisture forecasting model are displayed in Figure 7. The x-axis represents the time in hours, and the y-axis represents the moisture content in percentage. The red line represents the actual moisture content, and the green line represents the forecasted moisture content. The plot illustrates the ability of the LSTM model to accurately predict the moisture content of transformer oil-paper insulation over time. The plot shows a good alignment between the actual and forecasted values, indicating that the model has a high level of accuracy and reliability.

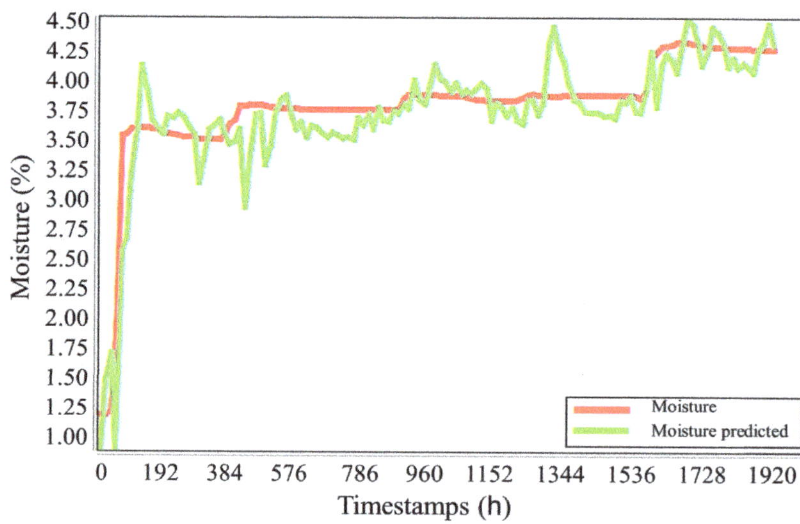

Figure 7. Results of the LSTM-based moisture forecasting model.

Table 1 summarizes the forecasting model's performance metrics using R^2, mean absolute error (MAE), mean squared error (MSE), root mean squared error (RMSE), and mean signed difference (MSD). These metrics are commonly used to evaluate the performance of forecasting models and provide a comprehensive evaluation of the model's accuracy and consistency. The R^2 measures how well the model fits the data; it ranges between

0 and 1, where one means that the model perfectly fits the data. MAE, MSE, and RMSE are measures of the error between the predicted and actual values. MSD measures the difference between the predicted and actual values that consider the difference's sign. The values in the table represent the average error or difference across all the predictions made by the model. An R-squared value of 0.899 indicates that the model fits the data well and explains a large proportion of the variance in the observed oil moisture levels. This value indicates that the LSTM model is able to explain 89.9% of the variance in the moisture content of transformer oil-paper insulation.

Table 1. Performance metrics of the moisture forecasting model.

R^2	MAE	MSE	RSME	MSD
0.899	0.166	0.058	0.241	0.060

The proposed model's MAE, MSE, and RMSE values were determined to be 0.166, 0.058, and 0.241, respectively, correspondingly showing a modest average difference between predicted and actual moisture levels. The proposed approach was also compared to two regularly used moisture forecasting models: a traditional linear regression model and a random forest (RF) model. In terms of accuracy, the suggested LSTM model outperformed the classic linear regression model and the RF model, with an R-squared value of 0.899 vs. 0.376 and 0.654, respectively. The LSTM model also outperformed linear regression and RF in terms of MAE, MSE, and RMSE, with values of 0.166, 0.058, and 0.241, respectively, versus 0.226, 0.172, and 0.414, respectively, for linear regression and 0.205, 0.151, and 0.387 for RF, respectively. Table 2 compares the proposed model's performance metrics with linear regression and the random forest model.

Table 2. Comparison of the performance metrics of the proposed model with linear regression and random forest models.

MODELS	R^2	MAE	MSE	RSME
LR	0.376	0.226	0.172	0.414
RF	0.654	0.205	0.151	0.387
LSTM	0.899	0.166	0.058	0.241

Table 3 provides an average comparison of actual moisture, moisture forecast, and percentage error for different samples for ten iterations. The performance of the proposed model was evaluated using percentage error, a widely used metric to measure the accuracy of forecasting models. The percentage errors for the proposed model were calculated using the formula: percentage error = (| actual value–predicted value | /actual value) × 100. The results showed that the percentage errors for the proposed model were 3.22, 3.13, 0.54, 0.52, and 0.48 for five different samples. The proposed model achieved an average percentage error of 1.57%. These results demonstrate the potential of the proposed model to provide more accurate predictions of moisture content in transformer oil-paper insulation.

Table 3. Comparison of actual moisture, moisture forecast, and percentage error.

Moisture (%) (Actual)	Moisture (%) (Forecast)	Absolute Error (Forecast)	Percentage Error (Forecast)
1.24	1.20	0.04	3.22
1.28	1.24	0.04	3.13
3.70	3.68	0.02	0.54
3.87	3.85	0.02	0.52
4.12	4.10	0.02	0.48

4. Discussion

The maintenance and longevity of power transformers are crucial for ensuring stability in energy systems. The moisture in transformer oil-paper insulation can cause deterioration and result in electrical failures. Therefore, to prevent such issues, it is crucial to forecast the moisture levels in the insulation accurately. In this regard, this study proposes a novel approach to moisture diagnosis that uses a hybrid DFDS-LSTM method. The hybrid method combines the benefits of deep feature selection, and long short-term memory techniques have the potential to improve the accuracy of moisture prediction. The proposed method was tested using a dataset of transformer oil moisture readings. In terms of accuracy and efficiency, the proposed method performed well with an R^2 value of 0.899, thus proving itself as a valuable tool for power system operators to monitor and manage the integrity of their transformer fleet. This study highlights the potential of a deep learning approach for accurately identifying transformer oil insulation moisture and contributes to developing effective prognostic health management systems for transformers.

In this study, the performance of long short-term memory (LSTM) networks was compared with linear regression and random forest models for forecasting transformer moisture. According to the findings, the LSTM model performed significantly better in performing accurate predictions than the linear regression or the random forest models. LSTM networks could capture long-term dependencies in the data by using a memory cell, which allows them to maintain a certain level of information for an extended period. Hence it is beneficial for transformer moisture forecasting, as the moisture content of the transformer oil-paper insulation is affected by both short-term and long-term factors, such as temperature, aging, and dielectric properties. In contrast, linear regression and random forest models are not intended to handle time series data and cannot capture long-term dependencies within the data [22]. In this investigation, the linear regression model failed to capture the non-linear relationship between input characteristics and moisture content. The random forest model captured the non-linear relationship between the input features and the moisture content, although it performed worse than the LSTM model. As a result of their ability to capture long-term dependencies in the data, LSTM networks are especially suited for transformer moisture forecasting tasks. Furthermore, their capacity to handle non-linear relationships between input features and output makes them a better alternative for transformer moisture predictions than linear regression and random forest models.

5. Conclusions

In this research, accelerated aging insulation samples were prepared in the laboratory to rapidly obtain the moisture-effected aging characteristics of the transformer insulation. Dielectric spectroscopy of oil-paper samples was obtained; however, moisture-sensitive features are complex to extract for accurate moisture forecasting. A deep learning method based on LSTM was proposed to forecast moisture variations inside the transformer tank through non-intrusive DFDS feature parameters. The proposed model was validated by collecting data from a real-life transformer, and the results showed that the proposed model was able to accurately predict moisture variations inside the transformer tank. The DFDS dataset was analyzed to extract moisture-sensitive features, and the LSTM-based deep learning method was proposed to forecast the moisture variation inside the transformer tank. The model was trained and validated using the DFDS dataset and the moisture measurements of the samples. The results of this study provide insight into the moisture variation inside the transformer oil-paper insulation and demonstrate the potential of the proposed LSTM-based model to provide accurate and non-intrusive moisture forecasting.

Overall, this research demonstrates the effectiveness of using LSTM-based deep learning for forecasting moisture variations in transformer oil-paper insulation. The proposed model can be used for prognostic monitoring of transformer oil-paper insulation and predicting the transformer's health status. This study can help prolong the transformer's life and reduce the need for unplanned repairs or replacements, eventually saving energy.

Furthermore, the results of this research can potentially improve the electrical grid's overall performance by reducing energy losses and minimizing the risk of unexpected outages.

Author Contributions: Conceptualization, A.V. (Aniket Vatsa) and A.S.H.; Data curation, A.V. (Aniket Vatsa), A.S.H. and V.B.; Formal analysis, V.P. and P.C.; Investigation, A.V. (Alexander Vinogradov) and P.C.; Methodology, A.V., A.S.H. and V.B.; Resources, V.B., A.V. (Alexander Vinogradov) and V.P.; Software, A.V. (Aniket Vatsa), V.B. and A.V. (Alexander Vinogradov); Supervision, A.S.H., V.B., A.V. (Alexander Vinogradov), V.P. and P.C.; Validation, A.S.H., V.P. and P.C.; Visualization, A.V. (Aniket Vatsa) and V.B.; Writing—original draft, A.V. (Aniket Vatsa), A.S.H., V.P. and P.C.; Writing—review & editing, A.V. (Aniket Vatsa), A.S.H., V.B. and A.V. (Alexander Vinogradov). All authors have read and agreed to the published version of the manuscript.

Funding: This research received no external funding.

Data Availability Statement: Data sharing is not applicable.

Conflicts of Interest: The authors declare no conflict of interest.

References

1. Li, A.; Yang, X.; Dong, H.; Xie, Z.; Yang, C. Machine learning-based sensor data modeling methods for power transformer PHM. *Sensors* **2018**, *18*, 4430. [CrossRef]
2. Fernandez, I. The Need for Experimental and Numerical' Analyses of Thermal Ageing in Power Transformers. *Energies* **2022**, *15*, 6393. [CrossRef]
3. Vatsa, A.; Hati, A.S. Depolarization Current Prediction of Transformers OPI System Affected from Detrapped Charge Using LSTM. *IEEE Trans. Instrum. Meas.* **2022**, *71*, 2511711. [CrossRef]
4. Liu, J.; Fan, X.; Zhang, Y.; Zheng, H.; Zhu, M. Quantitative evaluation for moisture content of cellulose insulation material in paper/oil system based on frequency dielectric modulus technique. *Cellulose* **2020**, *27*, 2343–2356. [CrossRef]
5. Zukowski, P.; Rogalski, P.; Kołtunowicz, T.N.; Kierczynski, K.; Zenker, M.; Pogrebnjak, A.D.; Kucera, M. DC and AC Tests of Moisture Electrical Pressboard Impregnated with Mineral Oil or Synthetic Ester—Determination of Water Status in Power Transformer Insulation. *Energies* **2022**, *15*, 2859. [CrossRef]
6. Liu, H.; Jia, H.; Liu, J.; Meng, F. Method for quantitative assessment of transformer oil-paper insulation non-uniform ageing parameters based on frequency domain dielectric response. *IET Sci. Meas. Technol.* **2022**, *16*, 118–129.
7. Hernandez, G.; Ramirez, A. Dielectric Response Model for Transformer Insulation Using Frequency Domain Spectroscopy and Vector Fitting. *Energies* **2022**, *15*, 2655. [CrossRef]
8. Rahman, O.; Islam, T.; Khera, N.; Khan, S.A. A novel application of the cross-capacitive sensor in real-time condition monitoring of transformer oil. *IEEE Trans. Instrum. Meas.* **2021**, *70*, 9513212. [CrossRef]
9. Kondalkar, V.V.; Ryu, G.; Lee, Y.; Lee, K. Development of highly sensitive and stable humidity sensor for real-time monitoring of dissolved moisture in transformer-insulating oil. *Sens. Actuators B Chem.* **2019**, *286*, 377–385. [CrossRef]
10. Jiang, J.; Wu, X.; Wang, Z.; Zhang, C.; Ma, G.; Li, X. Moisture content measurement in transformer oil using micro-nano fiber. *IEEE Trans. Dielectr. Electr. Insul.* **2020**, *27*, 1829–1836. [CrossRef]
11. Quinones, L.I.A.; Lozano-Moncada, C.A.; Montenegro, D.A.B. Machine learning for predictive maintenance scheduling of distribution transformers. *J. Qual. Maint. Eng.* 2022, *ahead-of-print*. [CrossRef]
12. de Souza, J.A.F.; Silva, M.M.; Rodrigues, S.G.; Santos, S.M. A forecasting model based on ARIMA and artificial neural networks for end–OF–life vehicles. *J. Environ. Manag.* **2022**, *318*, 115616. [CrossRef] [PubMed]
13. Zha, W.; Liu, Y.; Wan, Y.; Luo, R.; Li, D.; Yang, S.; Xu, Y. Forecasting monthly gas field production based on the CNN-LSTM model. *Energy* **2022**, *260*, 124889. [CrossRef]
14. Zaengl, W.S. Dielectric spectroscopy in time and frequency domain for HV power equipment. I. Theoretical considerations. *IEEE Electr. Insul. Mag.* **2003**, *19*, 5–19. [CrossRef]
15. Chakravorti, S.; Dey, D.; Chatterjee, B. *Recent Trends in the Condition Monitoring of Transformers*; Power Systems Springer: London, UK, 2013.
16. Smagulova, K.; James, A.P. A survey on LSTM memristive neural network architectures and applications. *Eur. Phys. J. Spec. Top.* **2019**, *228*, 2313–2324. [CrossRef]
17. Yang, L.; Zou, T.; Deng, B.; Zhang, H.; Mo, Y.; Peng, P. Assessment of oil-paper insulation aging using frequency domain spectroscopy and moisture equilibrium curves. *IEEE Access* **2019**, *7*, 45670–45678. [CrossRef]
18. Li, C.; Dong, Z.; Ding, L.; Petersen, H.; Qiu, Z.; Chen, G.; Prasad, D. Interpretable Memristive LSTM Network Design for Probabilistic Residential Load Forecasting. *IEEE Trans. Circuits Syst. I Regul. Pap.* **2022**, *69*, 2297–2310. [CrossRef]
19. Xuan, Y.; Yu, Y.; Wu, K. Prediction of short-term stock prices based on emd-lstm-csi neural network method. In Proceedings of the 2020 5th IEEE International Conference on Big Data Analytics (ICBDA), Xiamen, China, 8–11 May 2020; pp. 135–139.
20. Dubey, S.R.; Chakraborty, S.; Roy, S.K.; Mukherjee, S.; Singh, S.K.; Chaudhuri, B.B. diffGrad: An optimization method for convolutional neural networks. *IEEE Trans. Neural Netw. Learn. Syst.* **2019**, *31*, 4500–4511. [CrossRef] [PubMed]

21. Ning, C.; You, F. Optimization under uncertainty in the era of big data and deep learning: When machine learning meets mathematical programming. *Comput. Chem. Eng.* **2019**, *125*, 434–448. [CrossRef]
22. Fan, G.F.; Yu, M.; Dong, S.Q.; Yeh, Y.H.; Hong, W.C. Forecasting short-term electricity load using hybrid support vector regression with grey catastrophe and random forest modeling. *Util. Policy* **2021**, *73*, 101294. [CrossRef]

energies

MDPI

Article

A Fractional Order Controller for Sensorless Speed Control of an Induction Motor

Tayyaba Nosheen [1,2], Ahsan Ali [1], Muhammad Umar Chaudhry [3,*], Dmitry Nazarenko [4], Inam ul Hasan Shaikh [1], Vadim Bolshev [4,5,*], Muhammad Munwar Iqbal [6], Sohail Khalid [2] and Vladimir Panchenko [7]

1 Department of Electrical Engineering, University of Engineering and Technology, Taxila 47070, Pakistan
2 Department of Electrical Engineering, Riphah International University, Islamabad 45210, Pakistan
3 Department of Computer Science, MNS-University of Agriculture, Multan 66000, Pakistan
4 Laboratory of Intelligent Agricultural Machines and Complexes, Don State Technical University, 344000 Rostov-on-Don, Russia
5 Federal Scientific Agroengineering Center VIM, 109428 Moscow, Russia
6 Department of Computer Science, University of Engineering and Technology, Taxila 47070, Pakistan
7 Department of Theoretical and Applied Mechanics, Russian University of Transport, 127994 Moscow, Russia
* Correspondence: umar.chaudhry@mnsuam.edu.pk (M.U.C.); vadimbolshev@gmail.com (V.B.); Tel.: +92-336-6034688 (M.U.C.); +7-499-174-8595 (V.B.)

Abstract: Agriculture activities are completely dependent upon energy production worldwide. This research presents sensorless speed control of a three-phase induction motor aided with an extended Kalman filter (EKF). Although a proportional integral (PI) controller can ensure tracking of the rotor speed, a considerable magnitude of ripples is present in the torque generated by a motor. Adding a simple derivative to have a proportional integral derivative (PID) action can cause a further increase in ripple magnitude, as it allows the addition of high-frequency noise in the system. Therefore, a fractional-order-based PID control is presented. The proposed control scheme is applied in a closed loop with the system, and simulation results are compared with the PID controller. It is evident from the results that the fractional order control not only ensures 20 times faster tracking, but ripple magnitude in torque was also reduced by a factor of 50% compared to that while using PID and ensures the effectiveness of the proposed strategy.

Keywords: sensorless control; extended Kalman filter; fractional order control; fractional calculus; non-integer integral-differential equations

check for updates

Citation: Nosheen, T.; Ali, A.; Chaudhry, M.U.; Nazarenko, D.; Shaikh, I.u.H.; Bolshev, V.; Iqbal, M.M.; Khalid, S.; Panchenko, V. A Fractional Order Controller for Sensorless Speed Control of an Induction Motor. *Energies* **2023**, *16*, 1901. https://doi.org/10.3390/en16041901

Academic Editor: Frede Blaabjerg

Received: 17 January 2023
Revised: 2 February 2023
Accepted: 9 February 2023
Published: 14 February 2023

1. Introduction

The arrival of electric motors helped the world in moving towards automation. Electric motors, especially induction motors (IMs), play a vital role in the field of electrical engineering due to that the induced voltage in the rotor results in a rotor current without any physical contact with the rotor windings. Due to the wide range of industrial applications, speed control to date is an important aspect of IMs.

Induction motors are known for their robustness, reliability, and efficiency, making them well-suited for use in harsh and outdoor environments. IMs also require less maintenance compared to other types. Induction motors are widely used in the agricultural industry to power various types of equipment, such as pumps, conveyors, and fans. They are particularly useful in irrigation systems, where they are used to power large pumps that draw water from wells or rivers and distribute it to the crops. Aside from that, they are also used to power grain elevators, threshers, and other machinery used in crop processing [1,2]. Induction motors can be connected to the power take-off (PTO) shaft of a tractor to drive various types of equipment such as plows, harrows, and mowers.

2. Literature Review

Considering the importance in terms of their application in industry, a variety of control strategies have been proposed in the literature for speed control of the IMs. A brief survey of different control approaches namely, Scalar control, Direct self-control, and Direct torque control (DTC), has been presented in [3]. In this research, scalar and field-oriented control problems of IM are moderately solved by vector control techniques (VCTs). The control design approach for IMs being used in heating, air conditioning, and ventilation has been presented in [4]. Critical analysis of torque ripples, tracking speed, control algorithm complexity parameter variation, and switching loss is presented in [5]. Field-oriented control (FOC) for the field weakening region of the IM is presented in [6]. The FOC is used to provide a parameter dependent control of torque and flux of the IM which can then be used in speed control of the motor. The main limitation of this work comes from the dependency of the approach on system parameters. A slight change in the motor parameters can cause instability in the system. To achieve high switching frequency, Direct self-control (DSC) is presented in [7]. DSC is similar to direct torque control (DTC) except one needs to achieve DTC by controlling the rotor flux.

In Ref. [8], a high-efficiency speed control approach for IM is presented. The basic principle of this operation involves employing a switching table for the selection of the output voltage of the inverter. In [9], hysteresis controllers are designed for stator flux and torque control for IM. The limitations of the controllers proposed in [9] include the presence of high-frequency ripples and the ability to achieve reference tracking of motor speed for a relatively shorter range. To address the aforementioned issues, DTC [10,11] multilevel converters [12,13] and model predictive control (MPC) [14] have been presented for IMs. A cascaded free control structure to reduce the complexity of MPC is presented in [15].

The space vector modulation technique for direct torque control was introduced in 1992, and since then, various structures have been proposed in the literature [16]. A PI controller is used to generate a suitable voltage vector with torque variation. A stator flux-oriented control strategy is presented in [17,18]. Instead of a hysteresis controller for generating voltage components, two PI controllers are used, and a sensorless algorithm is applied to estimate flux torque. This approach is used to minimize the cost of the installation of sensors and cabling. In [19], a PI controller with two different tuning approaches is used: one is symmetric optimum, and the other is root locus. The simulation results are validated through experimentation. In [20], an indirect field-oriented control technique is implemented on the induction motor. A proportional integral derivative controller is designed by researchers, and the controller parameter is tuned by the numerical method.

Aside from speed control, various techniques are also presented in the literature for speed estimation. A comparative study based on non-linear estimators is presented in [21]. In Ref. [22], a sliding-mode observer is used to estimate the motor's states. The dynamic model of the motor is presented in a dual frame of reference. In this paper, the low-speed region is controlled without using the signal injection technique [23]. An adaptive controller with an accurate model is created. Adaptive feedforward control compensates for uncertainty and non-linear factors. A comparative analysis is presented with three control approaches. The presented control technique is complex in structure and cannot be implemented for low-cost applications. A high-order sliding mode control technique twinning with an adaptive observer is proposed in [24]. The adaptive observers are highly dependent on the motor parameters and are not stable [25,26]. They provide motor control without mechanical sensors. The main drawback of this approach is a static error that is proportional to speed. In Ref. [27], a novel observer is developed for the linear system based on fuzzy logic. The Taylor series linearizes the fifth-order non-linear model of an induction motor. The T-S fuzzy model and observer were presented in this paper. Exponential convergence of estimation errors is also achieved in this study. Experimental results validate speed tracking. In Ref. [28], the rotor position is estimated using an adaptive filter. Furthermore, a reduction in the complexity of the mathematical model of a motor is also presented.

In Ref. [29], a simple flux regulation method is presented as a solution for the induction motor's low speed and no speed issues. A constant switching frequency controller is proposed in this paper. This research reduces the flux drop issues due to hysteresis control and offers a wide speed range of control.

Contributions of the Proposed Work

The significant contributions of this work have been listed as follows:

1. Speed control of an induction motor by designing three fractional order controllers for flux, speed, and torque along with space vector modulation (SVM) for ripple reduction.
2. Reference speed tracking, which provides tracking with less rise time, settling time, and overshoot Speed tracking covers a wide range, e.g., low to high, high to low, and again high. The speed reversal tracking (anti-clockwise rotation) of the three-phase induction motor is also highlighted.
3. The state estimation concept for a sensor-free system is provided. The induction motor model is extended by adding two more states: speed and load torque. The EKF estimator is designed for sensorless control, which provides ideal filtering of noise and estimates states with minimal error.

The paper is organized as follows: Section 2 describes the induction motor model used in speed control. Section 3 illustrates the control strategy of the induction motor. Sections 4 and 5 demonstrate the EKF and SVM techniques used in research, respectively. Simulated results are discussed in Section 6, and a brief conclusion is presented in Section 7.

3. Three-Phase Induction Motor

In this research, a three-phase squirrel-cage induction motor is considered where stationary reference frames are employed. In this frame of reference, the rotor winding is fixed to the frame and thus appears stationary. Using [25,30], equations from (1)–(16) are achieved. The winding on the *d-q* axis is designated as *ds*, *dr*, *qs*, and *qr*. The direct and quadrature axes of a three-phase induction motor are shown in Figure 1. Voltage and flux linkage equations can be given as [25]:

$$V_{ds} = R_s i_{ds} + \dot{\psi}_{ds} - \psi_{qs}\omega_r \tag{1}$$

$$V_{qs} = R_s i_{qs} + \dot{\psi}_{qs} + \psi_{ds}\omega_r \tag{2}$$

$$V_{dr} = R_r i_{dr} + \dot{\psi}_{dr} \tag{3}$$

$$V_{qr} = R_r i_{qr} + \dot{\psi}_{qr} \tag{4}$$

Flux-linkage Equations:

$$\psi_{ds} = L_s i_{ds} + L_m i_{dr} \tag{5}$$

$$\psi_{qs} = L_s i_{qs} + L_m i_{qr} \tag{6}$$

$$\psi_{dr} = L_r i_{dr} + L_m i_{ds} \tag{7}$$

$$\psi_{qr} = L_r i_{qr} + L_m i_{qs} \tag{8}$$

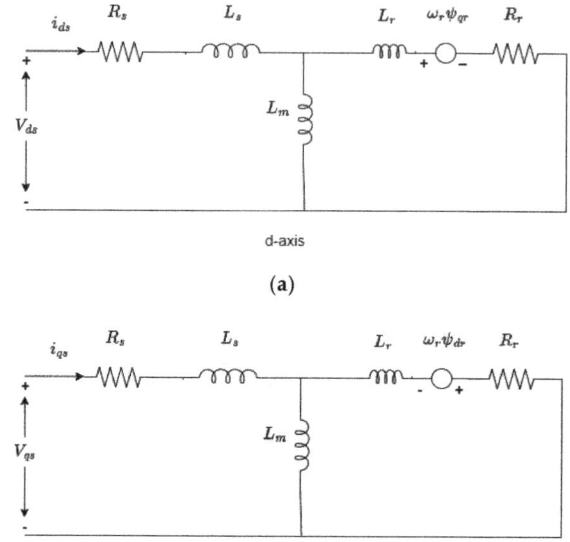

(a)

(b)

Figure 1. Axes of three−phase induction motor: (**a**) Direct, (**b**) Quadrature.

State Space Model

The time-dependent model of the induction motor can be given by Equations (9) and (10) where stator components, current, and flux are the state variables [25].

$$\dot{x}(t) = Ax(t) + Bu(t) \tag{9}$$

$$y(t) = Cx(t) \tag{10}$$

where,

$$x(t) = \begin{bmatrix} i_{s\alpha} \\ i_{s\beta} \\ \psi_{s\alpha} \\ \psi_{s\beta} \end{bmatrix},\ y(t) = \begin{bmatrix} i_{s\alpha} \\ i_{s\beta} \end{bmatrix}$$

$$A = \begin{bmatrix} a_1 & -\omega_r & a_2 & \omega_r a_3 \\ \omega_r & a_1 & -\omega_r a_3 & a_2 \\ -R_s & 0 & 0 & 0 \\ 0 & -R_s & 0 & 0 \end{bmatrix} \tag{11}$$

$$B = \begin{bmatrix} a_4 & 0 \\ 0 & a_4 \\ 1 & 0 \\ 0 & 1 \end{bmatrix},\ C = \begin{bmatrix} 1 & 0 \\ 0 & 1 \\ 0 & 0 \\ 0 & 0 \end{bmatrix}^T \tag{12}$$

where

$$a_1 = -\left(\frac{R_s}{\sigma L_s} + \frac{R_r}{\sigma L_r}\right), a_2 = \frac{R_s}{\sigma L_s L_r}, a_3 = \frac{1}{\sigma L_r},\ a_4 = \frac{1}{\sigma L_s}\ and\ \sigma = 1 - \frac{L_m^2}{L_s L_r}$$

The torque equation is defined as:

$$T_e = P(\psi_{s\alpha} i_{s\beta} - \psi_{s\beta} i_{s\alpha}) \tag{13}$$

The rotating speed of the stator magnetic field can be given as:

$$N_s = \frac{120 \times f}{n} \tag{14}$$

$$\dot{\omega}_r = \frac{1}{J}[T_e - T_L - f_r \omega_r] \tag{15}$$

A list of nomenclature used in this work is given in Table 1 at the beginning of the paper and the values of motor parameters are provided in Table 2.

Table 1. List of parameters.

Symbols	Description
$i_{s\alpha}$	Stator current along the alpha axis
$i_{s\beta}$	Stator current along the beta axis
i_{ds}	Stator current along the direct axis
i_{qs}	Stator current along the quadrature axis
$\psi_{s\alpha}$	Stator flux along the alpha axis
$\psi_{s\beta}$	Stator flux along the beta axis
ψ_{ds}	Stator flux along the direct axis
ψ_{qs}	Stator flux along the quadrature axis
V_{ds}	Stator voltages along the direct axis
V_{qs}	Stator Voltage along quadrature axis
i_{dr}	Rotor current along the direct axis
i_{qr}	Rotor current along the quadrature axis
ψ_{dr}	Rotor flux along the direct axis
ψ_{qr}	Rotor flux along the quadrature axis
V_{dr}	Rotor voltages along the direct axis
V_{qr}	Rotor voltages along quadrature axis
σ	Blondel's coefficient
T_e	Electromagnetic torque
T_L	Load torque
ω_r	Rotor speed
n	Number of poles
f	Frequency
V_{DC}	DC-bus voltage

Table 2. Motor parameters [25].

Parameter	Symbol	Value	Unit
Resistance of stator	R_s	6.75	Ω
Resistance of rotor	R_r	6.21	Ω
Inductance of stator	L_s	0.51	H
Inductance of rotor	L_r	0.5192	H
Mutual inductance	L_m	0.4957	H
Friction coefficient	f_r	0.002	Nm/rad
Inertia coefficient	J	0.01240	kg m^2
Poles	p	2	-
Power	P	1.1	kW

4. Control Strategy

Despite the arrival of many effective design methods in the control field, PID controllers are undeniably adopted in industrial settings. The primary reason for the excessive use of PID in the industry is its cost-benefit ratio. In this research, fractional order proportional integral derivative (FOPID) and PI controllers are used for speed, flux, and torque control. To apply the controllers in the closed loop, a stator flux-oriented control strategy is adopted in this work. Torque is aligned to the quadrature axis, while the stator flux vector is aligned along the direct axis, maintaining the quadrature component of stator flux at zero. The global control strategy is done by following the steps given below:

1. The estimated torque, speed, and flux signal are subtracted from the reference signal.
2. The difference between the reference and estimated signal (error signal) is then acting as input to the controller.
3. The reference signal is converted into (α, β) coordinate system. Using Clark transformation and the knowledge of line voltage, current, and voltage in a stationary frame of reference are obtained. Clark and Inverse Clark transformation matrices are given in (16) and (17).
4. After modulation, the signal is then fed into the inverter.

The control scheme is shown in Figure 2.

$$\begin{bmatrix} v_\alpha \\ v_\beta \end{bmatrix} = \begin{bmatrix} 0.66 & 0.33 & -0.33 \\ 0 & 0.57 & -0.57 \\ 0.33 & 0.33 & 0.33 \end{bmatrix} \begin{bmatrix} v_a \\ v_b \\ v_c \end{bmatrix} \tag{16}$$

Inverse Clark transformation is used to achieve three-phase voltages [30].

$$\begin{bmatrix} v_a \\ v_b \\ v_c \end{bmatrix} = \begin{bmatrix} cos\theta & -sin\theta & 1 \\ cos\left(\theta - \frac{2\pi}{3}\right) & -sin\left(\theta - \frac{2\pi}{3}\right) & 1 \\ cos\left(\theta + \frac{2\pi}{3}\right) & -sin\left(\theta + \frac{2\pi}{3}\right) & 1 \end{bmatrix} \begin{bmatrix} v_\alpha \\ v_\beta \end{bmatrix} \tag{17}$$

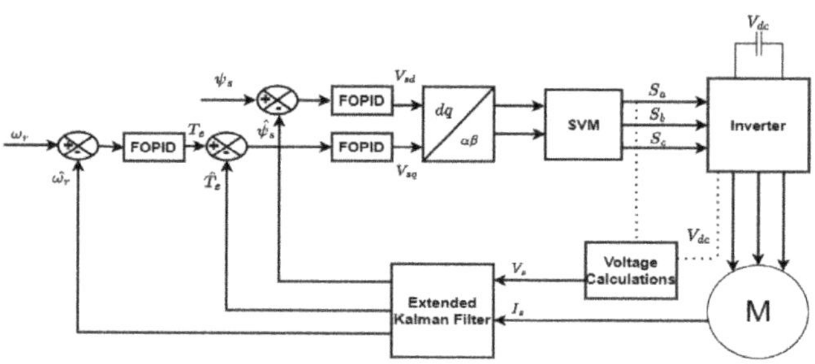

Figure 2. Proposed control Strategy using extended Kalman filter.

Fractional—Order Controller

Fractional order Derivatives are frequently used to deal with any order of integral or derivative. Among different kinds of definitions of derivatives and integrals, commonly used definitions are Grünwald—Letnikov, Riemann-Liouville, and Caputo. Using Euler's gamma function, the integer order is introduced by Riemann and Louisville [31], which defines the fractional order derivative as:

$${}_aD_t^\alpha f(t) = \frac{1}{\Gamma(n-\alpha)} \frac{d^n}{dt^n} \int_a^t \frac{f(\tau)}{(t-\tau)^{\alpha-n+1}} d\tau, n-1 < \alpha < n, \tag{18}$$

where a and t are the upper and lower bounds respectively, α is the order of derivative and the operator ${}_aD_t^\alpha$ is the fractional differentiation or integrals [31,32]. The Laplace transform technique is used to find the continuous time transfer function of integer-order systems. The Laplace transform of the signal for an nth-order derivative can be given by (19) [31].

$$L\{D^n x(t)\} = \int_0^\infty e^{-st} {}_0D_t^n x(t) dt = s^n X(s) - \sum_{k=0}^{m-1} s^k {}_0D_t^{n-k-1} x(t)|_{t=0} \tag{19}$$

where n has the range defined as $m-1 < n < m$ and m belongs to Z [33]. Oustaloup presented the concept of employing a fractional order controller for a dynamic system. The Oustaloup method is one of the real-time approximation techniques that uses a recursive distribution of poles and zeroes. The transfer function based on Oustaloup's method in the given frequency range $[\omega_l\ \omega_h]$ is given in (20).

$$s^v \approx k\prod_{n=1}^{N} \frac{1+\frac{s}{\omega_{z,n}}}{1+\frac{s}{\omega_{p,n}}} \tag{20}$$

In the approximation, order plays an important role in the performance of the approximating transfer function. The value of gain is trained in such a way that the overall gain of the transfer function is maintained at 1 rad/s. Gain and phase dynamics are distorted by low-order approximations. Higher orders of N can be utilized to eliminate ripples [34]. The approximate method will necessitate more processing power. In (20), the pole and zero frequencies are stated by the following set of equations [34]:

$$\omega_{z,1} = \omega_l\sqrt{\eta} \tag{21}$$

$$\omega_{p,n} = \omega_{z,n}\varepsilon, n = 1,\ldots,N \tag{22}$$

$$\omega_{z,n+1} = \omega_{p,n}\eta, n = 1,\ldots,N-1 \tag{23}$$

$$\varepsilon = \left(\frac{\omega_h}{\omega_l}\right)^{\frac{v}{N}} \tag{24}$$

$$\eta = \left(\frac{\omega_h}{\omega_l}\right)^{\frac{1-v}{N}} \tag{25}$$

The value of v plays an important role in approximation e.g., when it is less than 0, the equation behaves as an inverting equation and If $|v| > 1$, the estimate is not accurate enough. As a result, it is necessary to approximate δ in (26).

$$s^v = s^n S^\delta, n \in Z, \delta[0,1] \tag{26}$$

$$G_c(s) = K_p + K_i s^{-\lambda} + K_d s^\mu \tag{27}$$

Three parameters K_p, K_i, K_d and two orders μ, λ with non-integer values should be optimized while constructing a FOPID controller. Taking (μ, λ) = (1, 1), (1, 0), (0, 1), (0, 0), the classical controllers PID, PI, PD, and P are obtained. Figure 3a [35] shows the results obtained through the classical controller. These traditional varieties of PID controllers are all variations of the FOPID controller. Figure 3b [35] demonstrates how the fractional order controller expands upon the traditional PID controller and spreads from a point to a plane. This extension may provide and increase the controller's versatility in more precisely establishing control, objectives, and actual processes.

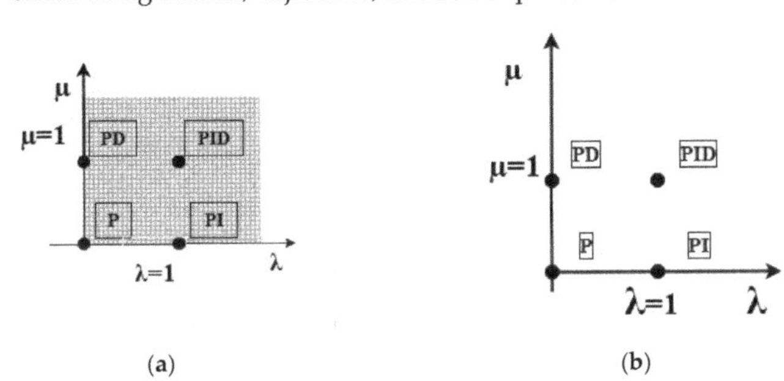

Figure 3. PID: (**a**) Classical order; (**b**) Fractional order.

5. Extended Kalman Filter (EKF)

In estimation theory, EKF is a non-linear filter used to estimate the non-linear states of systems. The Kalman filter algorithm is optimal in terms of minimizing the variance of the error estimation. It directly deals with system errors and treats them as noise. In the case of a three-phase induction motor, if speed is a state variable, then the state model becomes non-linear, so a suitable estimator for such non-linear systems is the EKF. Equations (28) and (29) express the generic non-linear model used for the state estimation process.

$$\frac{dx(t)}{dt} = f(x(t), u(t)) + w(t) \tag{28}$$

$$y = h(x(t)) + v(t) \tag{29}$$

where $w(t)$ and $v(t)$ denote noise and measurement noise, respectively.

EKF Algorithm

EKF algorithm is established to provide sensorless control that is required for state estimation. The implementation of the filter state-space model of the motor is required. After knowing state matrices, it is required to calculate the state transition matrix and prediction state. By minimizing the estimation error covariance, optimality of the estimated state is achieved. Initialization, prediction, and correction are involved in the EKF process to estimate torque, speed, and flux. The extended state variables vector of the induction motor can be defined as given in (30).

$$\begin{bmatrix} x_1(t) \\ x_2(t) \\ x_3(t) \\ x_4(t) \\ x_5(t) \\ x_6(t) \end{bmatrix} = \begin{bmatrix} i_{s\alpha} \\ i_{s\beta} \\ \psi_{s\alpha} \\ \psi_{s\beta} \\ \omega_r \\ T_L \end{bmatrix} \tag{30}$$

The filter uses the current estimation of inputs and states. f is the nonlinear function of the states and input. Input u is the alpha and beta components of stator voltage. Linearization is done by using (31). Function $F(t)$ and $H(t)$ are the Jacobians. $\hat{x}(t) = E(x(t))$ is the state space model whose states are estimated at t.

$$F(t) = \frac{\partial f}{\partial x}(\hat{x}(t), u(t)) \tag{31}$$

$$H(t) = \frac{\partial h}{\partial x}(\hat{x}(t)) = \begin{bmatrix} 1 & 0 & 0 & 0 & 0 & 0 \\ 0 & 1 & 0 & 0 & 0 & 0 \end{bmatrix} \tag{32}$$

Using F, the error covariance matrix is calculated as given in Equation (33). Initially, the value of the error covariance matrix P is set to zero. Matrices Q and R are defined in (34).

$$\dot{P}_c(t) = F(t)P_c(t) + P_c(t)F(t)^T - K(t)H(t)P_c(t) + Q(t) \tag{33}$$

where $K(t) = P_c(t)H(t)^T R(t)^{-1}$.

System parameter values and covariance matrix are very important and effective on the states estimation accuracy. Computational complexity avoided by covariance matrices of measurement and system noises is chosen in diagonal form. These matrices are obtained by considering the stochastic properties of noises [36,37]. The choice of R and Q matrix in EKF filter design is significant. For state estimation design of the Q matrix is very crucial as it affects the estimation error, and it is a weighting matrix. The literature proves that the solution to the Riccati equation for error covariance is a first-order approximation. So, when the system is linearized, then it is noisier than the non-linear system. So, tuning the Q matrix is very critical. From the stability property of EKF, if the value of the Q matrix is positive definite, then the Controllability Gramian of (F, Q) is bounded and also positive definite. This property for a non-linear system like our system helps in selecting the Q matrix. This results in less estimation error. From the literature on control, an increase in cost is inversely proportional to response speed. Q is tuned by adding a manual weight of less than 50, which results in a significant estimation error. After that, it is increased, which results in a decrease in error. Thus, this trial-and-error method is used to achieve the desired state estimation with minimum error. Matrices Q and R are covariance matrices of $w(t)$ and $v(t)$ which are given in Equation (34). Figure 4 shows the structure of extended Kalman filter.

$$R = \begin{bmatrix} 1 & 0 \\ 0 & 1 \end{bmatrix}, Q = \begin{bmatrix} \alpha & 0 & 0 & 0 & 0 & 0 \\ 0 & \alpha & 0 & 0 & 0 & 0 \\ 0 & 0 & \alpha & 0 & 0 & 0 \\ 0 & 0 & 0 & \alpha & 0 & 0 \\ 0 & 0 & 0 & 0 & \alpha_1 & 0 \\ 0 & 0 & 0 & 0 & 0 & \alpha_1 \end{bmatrix} \tag{34}$$

where $\alpha = 104$ and $\alpha_1 = 105$.

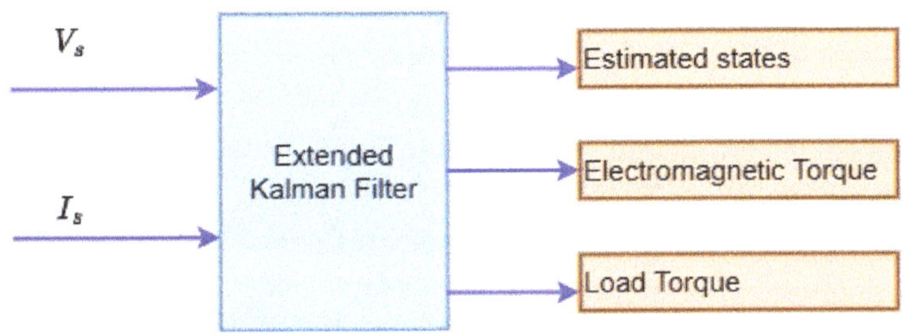

Figure 4. Structure of EKF.

6. Space Vector Modulation

A three-phase bridge inverter is used after modulation with eight switching states, and for all states, a corresponding voltage vector is generated. Figure 5 shows all switching states. For the three-phase, we have a revolving reference voltage vector that rotates in each sector. If this vector is in sector 1, voltage vector 1 and 2 is applied according to their application time by resolving it into their components. The rest of the time, the null/zero vector is applied, so the concept of the average voltage vector is applied in SVM over a sub-cycle Ts. Application time T1, T2, and T0 of the reference vector is shown below.

$$T_1 = \frac{T_s}{2V_{dc}}\left(\sqrt{6}V_{s\beta ref} - \sqrt{2}V_{s\alpha ref}\right) \tag{35}$$

$$T_2 = \sqrt{2}\frac{T_s}{V_{dc}}\left(V_{s\alpha ref}\right) \tag{36}$$

$$T_o = T_s - (T_1 + T_2) \tag{37}$$

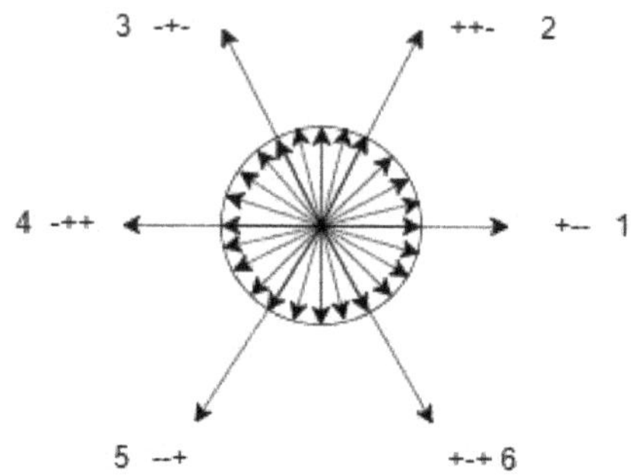

Figure 5. Space vector modulation.

7. Results and Discussion

The proposed control scheme has been implemented in MATLAB/Simulink and results have been presented in this section. The speed control test for the proposed controller has been done for four different cases. The reference inputs for all the cases have been discussed as follows:

1. Motor speed has been changed from 0 Revolutions per Minute (RPM) to 1000 RPM using a step function,
2. Motor speed is changed from 0 to 1000 RPM using a step signal and then it is suddenly reversed to −1000 RPM, i.e., it operated to 1000 RPM in opposite direction,
3. Motor speed is kept constant for 0.5 s and is then increased linearly with time to 487 RPM,
4. Motor speed is kept constant at 0 RPM for 1 s and then is increased to 500 RPM. It is kept constant at that speed for 1.5 s and is then suddenly increased to 1400 RPM. After keeping constant at 1400 RPM, it has been decreased to 0 RMP and is then increased to 500 RPM after being kept constant for 1 s.

The controller parameters for FOPID controllers are given in Table 3. It should be mentioned here that the comparison of results for the FOPID controller has been done with the PI controller proposed in [25].

Table 3. Controller parameters.

Controller Parameters for Speed	
K_p	2
K_i	0.01
λ	0.92
K_d	0.01
μ	0.85
Controller Parameters for Torque	
K_p	1000
K_i	100
λ	0.9
K_d	0
μ	0
Controller Parameters for Flux	
K_p	0.5
K_i	1
λ	0.9
K_d	0.01
μ	0.2

The controllers along with the EKF have been applied in the closed loop system as shown in Figure 2 and the simulation results have been obtained. The motor was provided a reference speed of 1000 RPM to track. The electromagnetic torque of the motor generated subjected to 1000 RPM reference speed is shown in Figure 6.

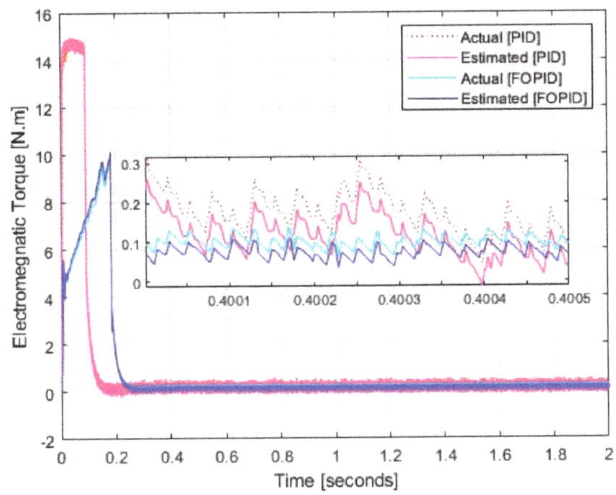

Figure 6. Electromagnetic torque.

It can be seen from Figure 6 that the torque generated by applying the PID controller has a peak of almost 15 Nm and then it decays down to almost 0. While the torque generated by the application of the FOPID controller has a peak of 10 Nm and then decays down to 0 Nm. It should be noted that the actual electromagnetic torque and the one estimated through EKF have the same profiles for both PID and FOPID controllers. It should also be noted that the ripple magnitude in electromagnetic torque produced with the FOPID controller has been reduced by almost one-third as compared to that with the PID controller. Figure 7 represents the stator flux magnitude, both actual and estimated through EKF, for the FOPID controller.

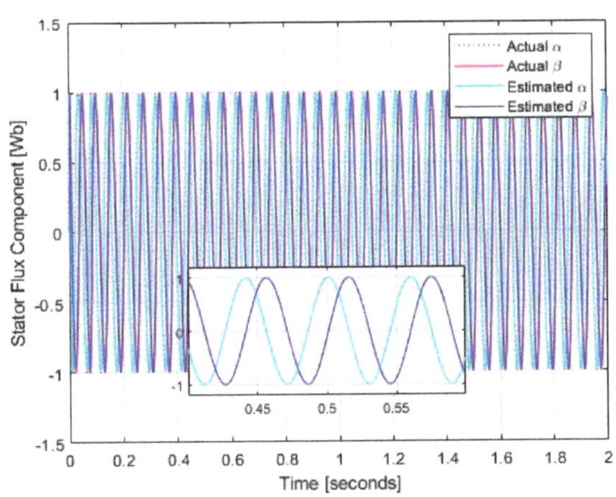

Figure 7. Stator flux components.

The stator flux magnitude for the induction motor in Figure 7 shows smooth sinusoidal variations for both flux components α and β. It should also be noted that the flux components for actual and estimated magnitudes are the same. The absence of ripples is evident in the effectiveness FOPID controller. Figure 8 shows the flux profile for the 1000 RPM reference speed of the motor. Moreover, it can be seen from Figure 7 that the flux has a faster tracking response with no ripples.

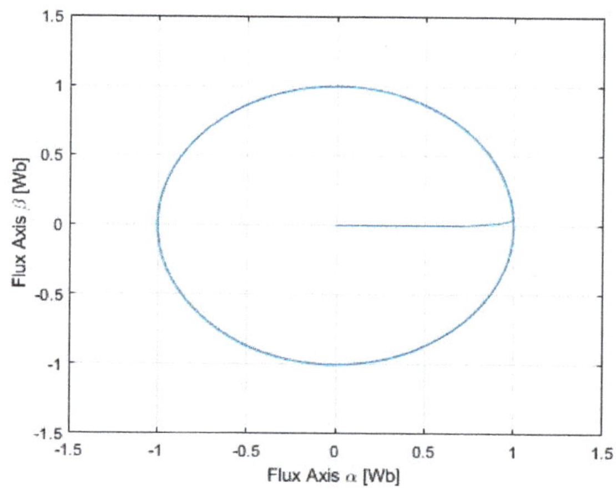

Figure 8. Stator flux circular trajectory.

The speed control for different reference trajectories with PID and FOPID controllers for speed variations discussed above is shown in Figure 9. At the outset, it can be stated that the proposed approach has been able to minimize high-frequency ripples from torque and stator flux of the IM as shown in Figures 6 and 7.

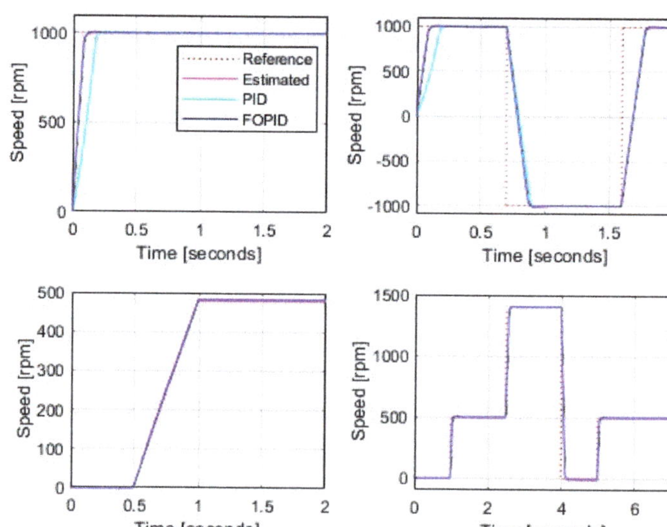

Figure 9. Rotor speed variations subjected to different reference trajectories.

It is evident from Figure 9 that the actual and estimated rotor speeds, using EKF with FOPID controller, of the induction motor are the same. However, the top two subplots of Figure 9 show that the FOPID controller ensures fast-tracking of the reference speed compared to the PID controller. It should also be noted that there is no overshoot in rotor speed for both PID and FOPID controllers. The increased time delay for the FOPID controller in speed tracking for the top-right subplot is because the reference trajectory has been designed such that rotor speed suddenly changes from 1000 RPM to −1000 RMP, i.e., the direction of movement of the rotor has been reversed while keeping the speed constant which is physically impossible. So, what the proposed control strategy does is that it reduces speed from 1000 RPM to 0 and then changes its direction and thereby reaching a speed of −1000 RPM. The speed tracking with PID and FOPID controllers for reference trajectories in the lower two cases of Figure 9 is the same.

The final comparison between the proposed control scheme and the sliding mode is presented in Table 4. The error between real and estimated states of torque, flux, and speed is presented. The error in flux and torque estimation is negligible, which shows that states are estimated with accuracy. Speed error is steadily decreasing with time and approaches zero. From Table 4, it can be seen that both techniques have 0% overshoot. However, when discussing the time delay parameter, it is clear from Table 4 that the presented technique shows its superiority.

Table 4. Comparative analysis of speed and Tracking error.

Speed Range	Time (s)	Time Delay (s)		Percent Overshoot		Comments
		FOPID	PID	FOPID	PID	
0–1000	0 to 0.7	0.1	0.1	0%	0%	Time delay same
1000 to 1000	0.7 to 1.6	0.2	0.27	0%	0%	Time delay of EKF is less (0.07 s)
−1000 to 1000	1.6 to 2	0.2	0.27	0%	0%	Time delay of EKF is less (0.07 s)
0 to 1000	0 to 0.7	0.07	0.1	0%	0%	Time delay of EKF is less (0.03 s)
100 to 600	0.7 to 1.4	0.02	0.45	0%	0%	Time delay of EKF is less (0.43 s)
600 to 1000	1.4 to 2	0.02	0.4	0%	0%	Time delay of EKF is less (0.38 s)
0 to 480	0 to 1			0%		
480 to 1430	1 to 2.5			0%		
1430 to 0	2.5 to 4	0	0	0%	0	This analysis is for a wide speed range. Less time delay for speed tracking highlights the fast performance of the proposed scheme.
0–1430	4 to 5			0%		
1430 to 480	5 to 7			0%		

Although a FOPID controller, due to two additional tunable parameters, provided more degree of freedom in control design, an integer order approximation of the fractional transfer function is required for hardware implementation. Higher integer order approximation results in higher accuracy and thus the computation complexity is increased. In the future, the supercapacitors may be designed to approximate the fractional derivatives to improve the hardware implementation complexity of the fractional order PID controllers.

A comparison of the proposed approach with others presented in the literature with respect to the time delay and percent overshoot has been given in Table 5. It is clear from Table 5 that the proposed approach not only supersedes the others in terms of a fast response, but it also has no overshoot in reference tracking of rotor speed.

Table 5. Comparison with existing techniques with no load.

Approach	Time Delay (s)	% Overshoot
Proportional integral (PI) + Slide Mode Observer (SMO) [25]	0.1	0%
Direct Torque Control (DTC) + Slide Mode Control (SMC) [38]	0.17	6.385%
PI + Space Vector Modulation (SVM)-DTC [39,40]	0.095	0%
Proposed technique (FOPID + extended Kalman filter (EKF))	0.07	0%

8. Conclusions

A non-integer differential-integral equations-based control strategy has been proposed for sensorless speed control of an induction motor. The idea is to control the rotor speed without its direct measurements at the output. So, an EKF has been designed to estimate the motor speed using torque and flux magnitudes. EKF successfully estimates the rotor speed and its comparison with the actual value shows negligible error. Since the torque magnitude when controlled through a PI controller has ripple magnitudes and derivative action can be a source of uncertainty due to high-frequency noise, fractional order derivatives, and integrals are used to design the controller. This non-integral order PID controller used along with SVM not only ensures speed tracking with negligible time delay, but it also minimizes the ripples to almost one-fourth magnitude as compared to that with PI controller. This work uses three different fractional order controllers for flux, torque, and speed control which requires their separate implementation when connected to hardware in a real-time system. This limitation may be eliminated in the future by designing a centralized controller to ensure torque, flux, and speed control and minimize the ripples magnitude at the same time.

Author Contributions: Conceptualization, T.N. and A.A.; methodology, A.A., M.U.C. and I.u.H.S.; software, T.N., D.N. and M.M.I.; validation, A.A., D.N., M.M.I. and V.B.; formal analysis, M.U.C., D.N., V.B. and V.P.; investigation, T.N., A.A., D.N. and S.K.; resources, A.A., M.U.C., I.u.H.S. and V.B.; data curation, T.N., A.A., D.N., I.u.H.S. and V.B.; writing—original draft preparation, T.N., A.A., I.u.H.S. and S.K.; writing—review and editing, D.N., V.B., M.M.I. and V.P.; visualization, T.N., M.M.I. and S.K.; supervision, A.A. and I.u.H.S.; project administration, M.U.C.; funding acquisition, M.U.C., V.B. and V.P. All authors have read and agreed to the published version of the manuscript.

Funding: This research received no external funding.

Data Availability Statement: No new data were created or analyzed in this study. Data sharing is not applicable to this article.

Conflicts of Interest: The authors declare no conflict of interest.

References

1. Kolhe, M.S.; Tapre, M.P.C. Condition Monitoring & Control of Induction Motors by using IoT Platform for Agriculture System. *Int. J. Eng. Res. Technol.* **2019**, *8*, 1043–1045.
2. Ramesh, A.; Kumar, M.S.; Sekhar, O.C. Interleaved boost converter fed with PV for induction motor/agricultural applications. *Int. J. Power Electron. Drive Syst.* **2016**, *7*, 835. [CrossRef]
3. Martins, C.A.; Carvalho, A.S. Technological trends in induction motor electrical drives. *IEEE Porto Power Tech Proc.* **2001**, *2*, 7.
4. Behrooz, F.; Mariun, N.; Marhaban, M.H.; Radzi, M.A.M.; Ramli, A.R. Review of control techniques for HVAC systems—Nonlinearity approaches based on Fuzzy cognitive maps. *Energies* **2018**, *11*, 495. [CrossRef]
5. El Ouanjli, N.; Derouich, A.; El Ghzizal, A. Modern improvement techniques of direct torque control for induction motor drives-a review. *Prot. Control Mod. Power Syst.* **2019**, *4*, 1–12. [CrossRef]
6. Vas, P.; Alakula, M. Field-oriented control of saturated induction machines. *IEEE Trans. Energy Convers.* **1990**, *5*, 218–224. [CrossRef]
7. Depenbrock, M. Direct self-control (DSC) of inverter fed induktion machine. *IEEE Power Electron. Spec. Conf.* **1987**, *3*, 632–641.
8. Takahashi, I.; Noguchi, T. A new quick-response and high-efficiency control strategy of an induction motor. *IEEE Trans. Ind. Appl.* **1986**, *5*, 820–827. [CrossRef]
9. Casadei, D.; Profumo, F.; Serra, G.; Tani, A. FOC and DTC: Two viable schemes for induction motors torque control. *IEEE Trans. Power Electron.* **2002**, *17*, 779–787. [CrossRef]
10. Lai, Y.-S.; Chen, J.-H. A new approach to direct torque control of induction motor drives for constant inverter switching frequency and torque ripple reduction. *IEEE Trans. Energy Convers.* **2001**, *16*, 220–227.
11. Lin, F.-J.; Wai, R.-J. Hybrid control using recurrent fuzzy neural network for linear induction motor servo drive. *IEEE Trans. Fuzzy Syst.* **2001**, *9*, 102–115.
12. Ahn, T.-C.; Kwon, Y.-W.; Hwang, H.-S.; Pedrycz, W. Design of neuro-fuzzy controller on DSP for real-time control of induction motors. In Proceedings of the 9th IFSA World Congress and 20th NAFIPS International Conference, Vancouver, BC, Canada, 25–28 July 2002; pp. 3038–3043.
13. Moallem, M.; Mirzaeian, B.; Mohammed, O.A.; Lucas, C. Multi-objective genetic-fuzzy optimal design of PI controller in the indirect field oriented control of an induction motor. *IEEE Trans. Magn.* **2001**, *37*, 3608–3612. [CrossRef]
14. Habibullah, M.; Lu, D.D.-C.; Xiao, D.; Rahman, M.F. A simplified finite-state predictive direct torque control for induction motor drive. *IEEE Trans. Ind. Electron.* **2016**, *63*, 3964–3975. [CrossRef]
15. Wróbel, K.; Serkies, P.; Szabat, K. Model predictive base direct speed control of induction motor drive—Continuous and finite set approaches. *Energies* **2020**, *13*, 1193. [CrossRef]

16. Ammar, A.; Bourek, A.; Benakcha, A. Modified load angle Direct Torque Control for sensorless induction motor using sliding mode flux observer. In Proceedings of the 4th International Conference on Electrical Engineering (ICEE), Boumerdes, Algeria, 13–15 December 2015; pp. 1–6.
17. Azcue-Puma, J.L.; Filho, A.J.S.; Ruppert, E. The fuzzy logic-based stator-flux-oriented direct torque control for three-phase asynchronous motor. *J. Control Autom. Electr. Syst.* **2014**, *25*, 46–54. [CrossRef]
18. Barambones, O.; Alkorta, P. Position control of the induction motor using an adaptive sliding-mode controller and observers. *IEEE Trans. Ind. Electron.* **2014**, *61*, 6556–6565. [CrossRef]
19. Zelechowski, M.; Kazmierkowski, M.P.; Blaabjerg, F. Controller design for direct torque controlled space vector modulated (DTC-SVM) induction motor drives. In Proceedings of the IEEE International Symposium on Industrial Electronics, Dubrovnik, Croatia, 20–23 June 2005; Volume 3, pp. 951–956.
20. Ferdiansyah, I.; Rusli, M.R.; Praharsena, B.; Toar, H.; Purwanto, E. Speed control of three phase induction motor using indirect field oriented control based on real-time control system. In Proceedings of the 10th International Conference on Information Technology and Electrical Engineering (ICITEE), Bali, Indonesia, 24–26 July 2018; pp. 438–442.
21. Manes, C.; Parasiliti, F.; Tursini, M. A comparative study of rotor flux estimation in induction motors with a nonlinear observer and the extended Kalman filter. In Proceedings of the IECON'94–20th Annual Conference of IEEE Industrial Electronics, Bologna, Italy, 5–9 September 1994; Volume 3, pp. 2149–2154.
22. Lascu, C.; Boldea, I.; Blaabjerg, F. Direct torque control of sensorless induction motor drives: A sliding-mode approach. *IEEE Trans. Ind. Appl.* **2004**, *40*, 582–590. [CrossRef]
23. Talla, J.; Leu, V.Q.; Šmídl, V.; Peroutka, Z. Adaptive speed control of induction motor drive with inaccurate model. *IEEE Trans. Ind. Electron.* **2018**, *65*, 8532–8542. [CrossRef]
24. Traoré, D.; Plestan, F.; Glumineau, A.; De Leon, J. Sensorless induction motor: High-order sliding-mode controller and adaptive interconnected observer. *IEEE Trans. Ind. Electron.* **2008**, *55*, 3818–3827. [CrossRef]
25. Ammar, A.; Bourek, A.; Benakcha, A. Sensorless SVM-direct torque control for induction motor drive using sliding mode observers. *J. Control Autom. Electr. Syst.* **2017**, *28*, 189–202. [CrossRef]
26. Zhou, M.; Cheng, S.; Feng, Y.; Xu, W.; Wang, L.; Cai, W. Full-Order Terminal Sliding-Mode-Based Sensorless Control of Induction Motor with Gain Adaptation. *IEEE J. Emerg. Sel. Top. Power Electron.* **2021**, *10*, 1978–1991. [CrossRef]
27. Liu, P.; Hung, C.-Y.; Chiu, C.-S.; Lian, K.-Y. Sensorless linear induction motor speed tracking using fuzzy observers. *IET Electr. Power Appl.* **2011**, *5*, 325–334. [CrossRef]
28. Cai, X.; Wang, Q.; Wang, Y.; Zhang, L. Research on a Variable-Leakage-Flux Permanent Magnet Motor Control System Based on an Adaptive Tracking Estimator. *Energies* **2023**, *16*, 587. [CrossRef]
29. Alsofyani, I.M.; Idris, N.R.N. Simple flux regulation for improving state estimation at very low and zero speed of a speed sensorless direct torque control of an induction motor. *IEEE Trans. Power Electron.* **2015**, *31*, 3027–3035. [CrossRef]
30. Chapman, S.J. *Electric Machinery Fundamentals*; McGraw-Hill: New York, NY, USA, 2004.
31. Zamani, M.; Karimi-Ghartemani, M.; Sadati, N.; Parniani, M. Design of a fractional order PID controller for an AVR using particle swarm optimization. *Control Eng. Pract.* **2009**, *17*, 1380–1387. [CrossRef]
32. Bingül, Z.; Karahan, O. Fractional PID controllers tuned by evolutionary algorithms for robot trajectory control. *Turk. J. Electr. Eng. Comput. Sci.* **2012**, *20*, 1123–1136. [CrossRef]
33. Rajasekhar, A.; Jatoth, R.K.; Abraham, A. Design of intelligent PID/PIλDμ speed controller for chopper fed DC motor drive using opposition based artificial bee colony algorithm. *Eng. Appl. Artif. Intell.* **2014**, *29*, 13–32. [CrossRef]
34. Podlubny, I. Fractional-order systems and PI/sup/spl lambda//D/sup/spl mu//-controllers. *IEEE Trans. Automat. Contr.* **1999**, *44*, 208–214. [CrossRef]
35. Bingul, Z.; Karahan, O. Comparison of PID and FOPID controllers tuned by PSO and ABC algorithms for unstable and integrating systems with time delay. *Optim. Control Appl. Methods* **2018**, *39*, 1431–1450. [CrossRef]
36. Vas, P. *Sensorless Vector and Direct Torque Control*; Oxford University Press: New York, NY, USA, 1998.
37. Ge, Q.; Feng, Z. Speed estimated for vector control of induction motor using reduced-order extended Kalman filter. In Proceedings of the Third International Power Electronics and Motion Control Conference, Beijing, China, 15–18 January 2000; pp. 138–142.
38. El Daoudi, S.; Lazrak, L.; Lafkih, M.A. Sliding mode approach applied to sensorless direct torque control of cage asynchronous motor via multi-level inverter. *Prot. Control Mod. Power Syst.* **2020**, *5*, 1–10. [CrossRef]
39. Ammar, A.; Benakcha, A.; Bourek, A. Closed loop torque SVM-DTC based on robust super twisting speed controller for induction motor drive with efficiency optimization. *Int. J. Hydrogen Energy* **2017**, *42*, 17940–17952. [CrossRef]
40. Premkumar, K.; Thamizhselvan, T.; Priya, M.V.; Carter, S.B.R.; Sivakumar, L.P. Fuzzy anti-windup pid controlled induction motor. *Int. J. Eng. Adv. Technol.* **2019**, *9*, 184–189. [CrossRef]

energies

MDPI

Article

Energy Efficient Received Signal Strength-Based Target Localization and Tracking Using Support Vector Regression

Jahir Pasha Molla [1], Dharmesh Dhabliya [2], Satish R. Jondhale [3], Sivakumar Sabapathy Arumugam [4], Anand Singh Rajawat [5], S. B. Goyal [6,*], Maria Simona Raboaca [7,*], Traian Candin Mihaltan [8], Chaman Verma [9] and George Suciu [10,*]

1 Department of Computer Science and Engineering, G. Pullaiah College of Engineering and Technology (GPCET), Kurnool 518002, India
2 Department of IT, Vishwakarma Institute of Information Technology, Pune 411048, India
3 Electronics and Telecommunication Department, Amrutvahini College of Engineering, Sangamner 422608, India
4 Department of ECE, Dr. N.G.P. Institute of Technology, Coimbatore 641048, India
5 School of Computer Science and Engineering, Sandip University, Nashik 422213, India
6 Faculty of Information Technology, City University, Petaling Jaya 46100, Malaysia
7 ICSI Energy Department, National Research and Development Institute for Cryogenics and Isotopic Technologies, 240050 Ramnicu Valcea, Romania
8 Faculty of Building Services, Technical University of Cluj-Napoca, 40033 Cluj-Napoca, Romania
9 Faculty of Informatics, University of Eötvös Loránd, 1053 Budapest, Hungary
10 R&D Department Beia Consult International, 041386 Bucharest, Romania
* Correspondence: sb.goyal@city.edu.my (S.B.G.); simona.raboaca@icsi.ro (M.S.R.); george@beia.ro (G.S.)

Abstract: The unpredictable noise in received signal strength indicator (RSSI) measurements in indoor environments practically causes very high estimation errors in target localization. Dealing with high noise in RSSI measurements and ensuring high target-localization accuracy with RSSI-based localization systems is a very popular research trend nowadays. This paper proposed two range-free target-localization schemes in wireless sensor networks (WSN) for an indoor setup: first with a plain support vector regression (SVR)-based model and second with the fusion of SVR and kalman filter (KF). The fusion-based model is named as the SVR+KF algorithm. The proposed localization solutions do not require computing distances using field measurements; rather, they need only three RSSI measurements to locate the mobile target. This paper also discussed the energy consumption associated with traditional Trilateration and the proposed SVR-based target-localization approaches. The impact of four kernel functions, namely, linear, sigmoid, RBF, and polynomial were evaluated with the proposed SVR-based schemes on the target-localization accuracy. The simulation results showed that the proposed schemes with linear and polynomial kernel functions were highly superior to trilateration-based schemes.

Keywords: received signal strength indicator (RSSI); trilateration; indoor localization; kalman filter (KF); support vector regression (SVR); generalized regression neural network (GRNN)

check for updates

Citation: Molla, J.P.; Dhabliya, D.; Jondhale, S.R.; Arumugam, S.S.; Rajawat, A.S.; Goyal, S.B.; Raboaca, M.S.; Mihaltan, T.C.; Verma, C.; Suciu, G. Energy Efficient Received Signal Strength-Based Target Localization and Tracking Using Support Vector Regression. *Energies* **2023**, *16*, 555. https://doi.org/10.3390/en16010555

Academic Editors: Pandian Vasant, Vladimir Panchenko, Vadim Bolshev, Igor S. Litvinchev, Prasun Chakrabarti and Nallapaneni Manoj Kumar

Received: 13 November 2022
Revised: 5 December 2022
Accepted: 26 December 2022
Published: 3 January 2023

1. Introduction

Moving object localization and tracking (L&T) is one of the most important research aspects for the success of various location-based-services (LBS) [1–3]. LBS can largely uplift the quality of life of those around us through numerous applications. For example, a rider can rent a bike with the help of a mobile app in a bike-sharing service and once their use is over they may leave it anywhere for the next user. In this application, the exact positions of all bikes in the service are utilized by interested customers to check for the nearest available bike. Nowadays, a smart-watch can very easily provide services such as person tracking, activity monitoring, and emergency messages. It has also been witnessed in the retail sector, insofar as the knowledge of exact positions can be utilized to raise business profits by identifying customer positions and guiding them towards specific products according to their interests in the mall. The people's locations at airports, metros, and rail stations can be used to study passenger statistics to provide the required signaling to them. Although GPS can provide locations, their indoor localization accuracy is limited because of the unavailability of satellite signals. Due to the capability of ubiquitous computing and smart sensing, the wireless sensor network (WSN) can replace GPS for indoor scenarios. In spite of other popular wireless technological options, the WSN is widely preferred for indoor L&T due to the low cost and low power consumption involved. The WSN is basically a network of number of sensor nodes that can cooperatively sense and transmit the parameters of interest from the surrounding RF environment to a base station at which the processing of the sensed data is possible [4].

Although a widely used metric for target L&T, RSSI field measurements are often very noisy and highly fluctuating in nature as indoor RF environments are generally very complex [5,6]. These measurements generally suffer from indoor interference, multi-path fading, noise, and varying obstacles That is why RSS-based target L&T systems often suffer from high localization errors. Due to its simplicity in implementation, trilateration is very popular for target L&T [7,8]. Although trilateration-based localization has simplicity in its implementation, it produces poor localization accuracy due to error uncertainty in RSSI measurements or the dynamic indoor environment. The localization accuracy generally suffers due to the fluctuating nature of RSS measurements. In more dynamic indoor environments involving reflection, interference, and obstacles, the machine learning (ML)-based RF fingerprinting techniques can be more useful than trilateration in the context of localization accuracy [9,10]. Considering the adaptive nature of ML techniques, they can easily eliminate the need for unnecessary system redesign. For instance, Support vector machine (SVM) has global optimality, very high data fitting accuracy, and fewer hyper parameters [11–13]. Due to high result generalization ability, SVM has also gained wide range preference in regression problems (wherein it is called as Support Vector Regression (SVR)). Compared with many popular ML models, SVR shows better forecasts in indoor target localization [10]. The SVM has four popular kernel functions, namely, linear, sigmoid, RBF, and polynomial. Thus, the proposed SVR-based system must be tested with these four kernel functions. One more important aspect in need of due consideration in the target-localization problem is the total energy consumed during the target-localization process [14]. Energy consumption during RF communication between two WSN nodes basically has two important components, namely, the energy consumed during RF propagation through the radio channel and the energy consumed in the WSN node hardware. This energy consumption is always directly proportional to length of the transmitted packet. While formulating the solution for the indoor target-localization problem, multiple RSSI measurements are generally taken into account to minimize the error in location estimation to compensate for the effect of the dynamicity in the RSSI measurement noise. However, due to the involvement of multiple RSSI measurements, the total energy consumption also increases. Thus, the proposed indoor localization approach must also be a low-energy-consumption solution. In this paper, we evaluated the impact of linear, sigmoid, RBF, and polynomial kernel functions on the target-localization accuracy as well as noted the energy consumption involved during localization in case I to case IV, respectively. The key outcomes of the research are as follows:

- An SVR-based L&T model fed with RSSI measurements was proposed to solve the problem of dynamicity in RSSI measurements as well as indoor environments, and it was compared with a well-known trilateration-based L&T scheme for the same RSSI measurements through rigorous localization accuracy simulations. Here, the trilateration and the proposed SVR-based scheme were fed with six and three RSSI measurements, respectively. The energy consumption during the target L&T for these two approaches were also compared.
- Further, the target location estimations obtained using the proposed SVR scheme were run through a standard Kalman Filter (KF) for further refinement, and named

as SVR+KF. The proposed SVR+KF framework was evaluated against trilateration and plain SVR-based schemes. Out of these three schemes, the SVR+KF-based scheme provided the lowest error in estimating the target location.

- We also tested the impact of the kernel function on target-tracking accuracy with the proposed SVR+KF algorithm. In this work, we tested four popular SVM kernel functions, namely, linear, sigmoid, RBF, and polynomial, during simulations in case I to case IV, respectively. In the target motion in all of these cases, the target was assumed to have high variation in the target velocity during its motion, and high maneuverability in trajectory. The noise in the RSSI measurements was kept the same for all four cases. The simulation results showed that the fusion of SVR and KF (i.e., the SVR+KF localization scheme) was highly accurate, consistent, and reliable in estimating target locations with the four considered types of kernels.

The rest of this paper is organized as follows: Section 2 covers the recent RSSI-based target-localization models, followed by a discussion on the proposed SVR-based target-localization model in Section 3. The system assumptions and design and the obtained results with the proposed L&T scheme are given in Sections 4 and 5, respectively. The research findings are then summarized in Section 6.

2. Related Work

The indoor target L&T schemes fed with RSSI measurements can be broadly categorized into two categories: ML-based methods and filter-based methods. The ML-based methods generally utilize supervised learning principles through RF fingerprinting. The popular ML-based L&T solutions in the literature are radial basis function (RBF), k-Nearest Neighbor (KNN), extreme learning machine (ELM), multilayer perceptron (MLP), recurrent Neural Network (RNN), Convolutional Neural Network (CNN), back propagation neural network (BPNN), and support vector machine (SVM). Once these models were trained offline with a dataset containing RSSI values and target locations, they were tested with random RSSI measurements in the online location estimation step. The error in the RSSI measurements is generally certain for almost any kind of indoor environment. The important reason behind this is the presence of household appliances, presence of electrical systems, and different kinds of obstacles between the transmitter and receiver nodes. Therefore, maintaining the line-of-sight (LOS) condition along the signal path between transmitter and receiver is impossible because of signal reflection, fading, and multi-path propagation. That means the mitigation of errors in the RSSI measurements is impossible. However, by adopting advanced ML-based signal-processing techniques, the target location estimation result can be improved. The ranging error can be mitigated with the help of a fuzzy-based obstacle identification and mitigation technique in the IR-UWB-based system [15,16]. The CNN-based target-localization scheme with RSSI measurements as inputs was proposed in ref. [17]. Here, the authors were successful in shifting the complexity of the online estimation stage to an offline training stage. The proposed scheme yielded 2 m localization accuracy. Here, thousands of RSSI fingerprints with entries for a 12.5 m $\times$ 10 m area were utilized for localization using the deployed APs. The average localization errors obtained with the proposed fingerprint-based approach were 4.1145 m, 4.1681 m, and 3.9118 m by utilizing SVM, KNN, and CNN-based schemes, respectively. The major drawback with the CNN-enabled target L&T schemes is the requirement of fine-tuning the hyper-parameters of CNN, namely, the activation function, threshold, and learning rate, and this is a very time-consuming task. This makes CNN accurate for specific indoor conditions, but less accurate for other indoor setups. In ref. [18], the authors proposed a kernel ELM (K-ELM)-based target L&T using 68,500 RSSI measurements obtained from an indoor area of 32 m $\times$ 16 m with eight sensor nodes. The proposed K-ELM-based scheme was compared with KNN, Bayesian, ELM, and online sequential ELM (OS-ELM) schemes, and it was found that the proposed scheme yielded 8.125 m accuracy, which is quite high against the rest of the other considered techniques for same indoor setup. The authors also used BPNN for target L&T, but it involved the need of a large number of iterations for converging to the optimum solution [19].

In ref. [20], the authors proposed a SVM-based L&T scheme, which is based on the assumption of advance knowledge of the node connectivity in the network and anchor node positions. They built an SVM classification model which utilizes RSSI field measurements collected by anchor nodes to be utilized for location estimation. This scheme was found to be reliable only for WSN with densely distributed nodes. In ref. [21], the authors computed the upper bound of the localization error for SVM-based L&T, using which the localization accuracy was improved using mass-spring-based optimization. In ref. [11], the authors presented a multi-class classifier based on SVM for L&T. It utilizes RSSI measurements obtained from a real-time environment of a hospital, and a laboratory building for training the proposed SVM classifier. In ref. [22], authors proposed a target L&T model using features of the channel state and RSSI. Herein, principal component analysis (PCA) was initially used for dimension reduction, and then SVM was used to obtain the target locations to obtain an accuracy in the range of 1 m. In our previous work [12], two range-free RSSI-based localization schemes, namely, SVR and SVR+KF were used. In this work, a linear kernel function with the proposed support vector regression (SVR) architecture was used to solve indoor target L&T. Unlike the work in ref. [12], in the proposed research work, the impacts of all the four popular kernel functions were tested with the SVR architecture, and we also attempted to find the energy consumption during the target L&T. The authors in ref. [13] proposed a least-squares-based SVR (LSSVR) to deal with dynamicity in RSSI measurements for a target L&T. The idea used here is to remove older values once new RSSI values are available in the queue. The proposed LSSVR scheme yielded an improvement of 21.82% without parameter optimization and of 11.70% with parameter optimization in localization accuracy.

In the filter-based target L&T, state-estimation techniques, such as KF and Particle Filter (PF), are major schemes, which involve two steps: prediction and measurement. The work in refs. [23,24] presented online semi-supervised SVR (OSS-SVR)-based localization to reduce the required amount of labeled data in the training set. Further, the proposed OSS-SVR results were fused with KF. It was found that the proposed OSS-SVR scheme was robust enough to in terms of the fluctuating system noise and needed a significantly smaller amount of labeled data during training. In our previous work [25], trilateration-based estimates were applied as inputs to KF for the tracking of mobile targets in WSN to present two range-based algorithms: RSSI+KF and RSSI+UKF. In this work, the proposed combination were evaluated for uncertainties in terms of RSSI noise, impact of variation in anchor density, and abrupt variation in target velocity. The results obtained through simulation experiments confirmed the efficacy of both presented algorithms in spite of RF environmental dynamicity. However, due to the need of frequently computing distances between transmitters and receivers, although the proposed algorithms showed localization errors below 1 m, it has large computational complexity as compared with other range-free localization solutions. The GRNN estimates were fed to KF to present range-free schemes for the target L&T in WSN [26–28]. The proposed algorithms GRNN+KF and GRNN+UKF successfully deal with RSSI noise uncertainty. Here, the proposed GRNN model was trained with only four RSSI measurements and the corresponding target locations obtained for any random time duration for the given indoor environment. Then, the GRNN-based location estimates obtained were supplied to KF and UKF to refine these further.

3. SVR for the Target L&T

SVM is a ML model which is based on supervised learning with several unique features as compared with other ML models. It can be used for classification (SVC) as well as regression [12]. The SVR has the capability to capture highly nonlinear relationships in the input-feature space and its computational complexity is not dependent on the dimensionality of the input space. It is capable of highly accurate prediction along with having a decent generalization ability. Due to all these advantages, it may be used for target

L&T. Here, the proposed SVR model was trained using 120 sets of RSSI field measurements and the corresponding target locations in the offline stage (See Figure 1). Once trained, any input vector of real-time RSSI measurements can be then applied to it to obtain the corresponding target-location estimate (online target-location estimation stage). In the background, the SVR architecture searches for similar RSSI input vectors from the training set to look for the closest possible match for the RSSI vector from the training set, and, based on that, it discerns the corresponding target-location estimate.

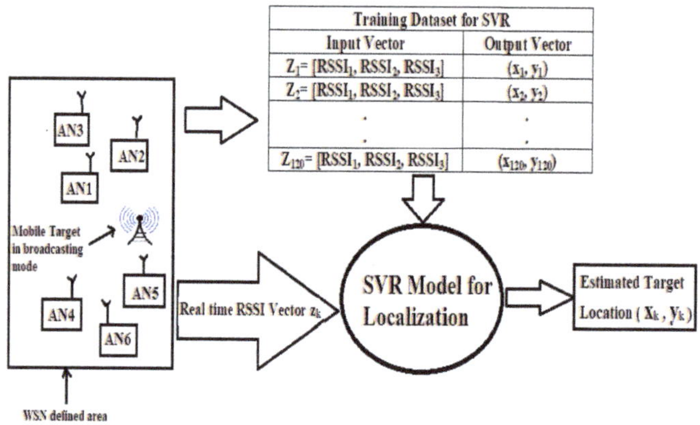

Figure 1. System block diagram for the proposed SVR-based target-localization scheme.

The simulated RSSI measurements considered in this work are based on a logarithmic shadowing model with the following mathematical equation [26,27]:

$$z_{\updownarrow j,k} = P_r(d_0) - 10n\log(d_{lj,k}/d_0) + X_\sigma, \tag{1}$$

where,
$(z_{\updownarrow j,k})$—RSSI received at the node $N_\updownarrow$ with coordinates $(x_{\updownarrow k}, y_{\updownarrow k})$ at time k,
η—Path loss exponent,
$P_r(d_0)$—RSSI measurement at the distance of d_0,
X_σ—Normal random variable representing the noise in RSSI.
The SVR model can be formulated by Equation (2) [21]:

$$F(z) = w^T z + b \tag{2}$$

where, b, and w are the SVR coefficients, and z is any given RSSI input vector. The optimized model corresponding to Equation (2) is given below [12]:

$$Minimize\frac{1}{2} \parallel w \parallel^2 + C\sum_{i=1}^{N}(\xi_i + \xi_i{}^*) \ \ subject\ to \begin{Bmatrix} F(z) - y_i \leq \varepsilon + \xi_i{}^* \\ y_i - F(z) \leq \varepsilon + \xi_i \\ \xi_i, \xi_i{}^* \geq 0, i = 1, 2, \ldots, N \end{Bmatrix} \tag{3}$$

where,
C—Regularization factor. Its default value ($C = 1$) is used here,
ε—Insensitive loss error function,
ξ_i, $\xi_i{}^*$—Upper and lower Slack variables of SVR.
The default values of γ and ε were used in this work, and were 0.01 and 0.001, respectively. To minimize Equation (2) into Equation (3), the regression function used is given below by Equation (4) [12]:

$$f(z) = \sum_{i=1}^{N}(\alpha_i^* - \alpha_i)K(z, z_i) + B, \tag{4}$$

where,

B—bias value,

$K(z, z_i)$—Kernel function,

$\alpha_i^*, \alpha_i \geq 0$—Lagrange multipliers.

Different kernel functions can be utilized to solve the target L&T problem with SVR [10,20]. In this work, we tested the SVR architecture for target-localization and tracking using four popular kernel functions, as given below.

(i) Linear Kernel

$$k(z, z_i) = z_i^T \cdot z \quad (5)$$

(ii) Sigmoid Kernel

$$k(z, z_i) = \tanh(\gamma(z_i^T \cdot z) + \beta) \quad (6)$$

where γ and β are constants. Here, as in ref. [21], we used $\gamma = 1/17$, $\beta = 0$. β is the slope parameter, which varies from 0 to ∞, yielding a straight line and step function, respectively. Thus, by varying the value of β, the slope of the sigmoid functions can be varied. As this research work aimed to estimate target location (regression problem), β was set to 0.

(iii) RBF Kernel

$$k(z, z_i) = \exp(-\gamma \| z - z_i \|^2), \quad (7)$$

(iv) Polynomial Kernel

$$k(z, z_i) = \left(\gamma(z_i^T \cdot z) + c\right)^d \quad (8)$$

where d is the degree of the polynomial and γ and c are the polynomial kernel constants. Here, as in Ref. [21], we used $\gamma = 1/17$, $c = 0$, and $d = 3$

4. System Design and Assumptions of the Proposed SVR-Based L&T System

In the presented work, an WSN area of 100 m × 100 m is considered with the motion of one target and six stationary anchor nodes (AN's) as shown in Figure 2 and Table 1. Out of all the six deployed AN's, any three AN's are required to locate a moving target using the proposed SVR and SVR+KF location estimation models. Here, RSSI measurements from AN1 to AN3 were given to the proposed schemes as inputs, whereas the measurements from all AN's were given to the trilateration-based localization scheme. The RSSI measurements obtained from the six AN's were denoted as $RSSI_1$ to $RSSI_6$. The key simulation parameters for this study are given in Table 2.

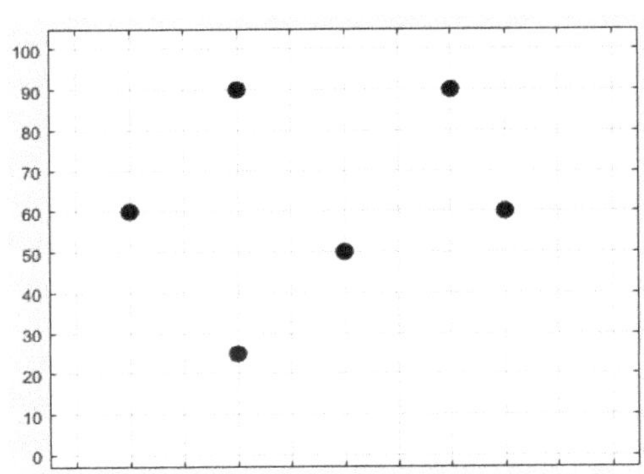

Figure 2. Plot of the location of the AN's in the considered WSN-defined area.

Table 1. Location of the ANs in the considered WSN-defined area.

Anchor Node Number	2-D Location	Anchor Node Number	2-D Location
AN1	(30, 25)	AN4	(30, 90)
AN2	(10, 60)	AN5	(80, 60)
AN3	(50, 50)	AN6	(70, 90)

Table 2. Simulation parameters.

Parameter	Value
Initial Target State X_0 at k = 0	(12, 16)
receiver and transmitter antenna gains	1 dB
AN communication radius	30 m
Transmission power	1 mW
Path Loss Exponent η	3
Discretization time step dt	1 s
X_σ	~N(3, 1)
Sigmoid Kernel Function Constant γ	1/17
Sigmoid Kernel Function Constant β	0
Polynomial Kernel Function Constant γ	1/17
Polynomial Kernel Function Constant c	0
Degree of the polynomial for Polynomial Kernel Function d	3

The input vector (X_k) for the proposed SVR-based schemes at a specific time instance k for each target location during its motion can be formulated as follows:

$$X_k = [RSSI_1, RSSI_2, RSSI_3],\ k = 1, 2, \ldots, 120 \tag{9}$$

Let us consider x_k and y_k as the target locations, $\dot{x}_k$ and $\dot{y}_k$ as the velocities in x and y directions, respectively, at time k^{th}. They are given as follows:

$$x_k = x_{k-1} + \dot{x}_k\, dt\,, \tag{10}$$

$$y_k = y_{k-1} + \dot{y}_k\, dt\,, \tag{11}$$

where $dt = k - (k-1)$ and is taken as 1 s here.

The target velocities during its motion for 40 locations are defined below using Equation (12) to Equation (15) (See Figure 3).

$$\dot{x}_k = 2,\ \dot{y}_k = 5,\ \text{for } 0 < k < 9 \text{ s}, \tag{12}$$

$$\dot{x}_k = 5,\ \dot{y}_k = 2,\ \text{for } 9 \le k \le 15 \text{ s}, \tag{13}$$

$$\dot{x}_k = 0,\ \dot{y}_k = 0,\ \text{for } 16 \le k \le 17 \text{ s}, \tag{14}$$

$$\dot{x}_k = 2,\ \dot{y}_k = -3,\ \text{for } 18 \le k \le 35 \text{ s}. \tag{15}$$

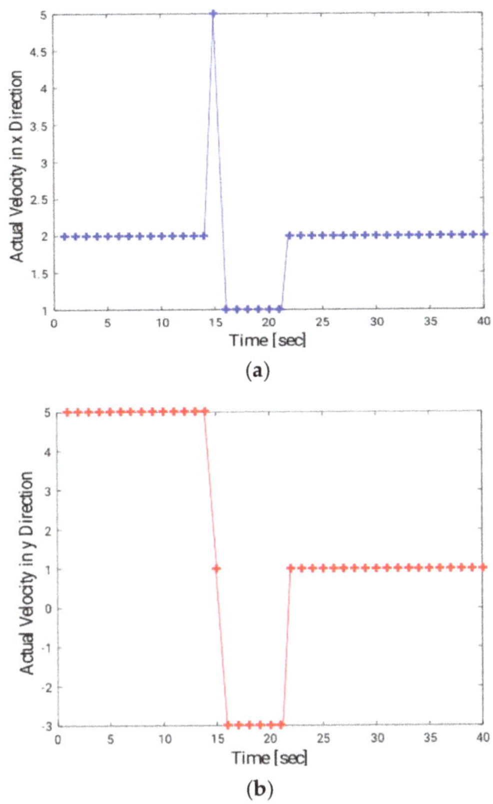

Figure 3. (**a**) Target velocity in x direction, (**b**) target velocity in y direction.

The performance evaluation parameters used in this work were root mean square error (RMSE) and average localization error as given below in Equation (16) to Equation (19). All the three RMSE's and localization error ideally must be as low as possible.

$$Average\ Localization\ Error = \frac{1}{T}\sum_{k=1}^{T}\frac{(\hat{x}_k - x_k) + (\hat{y}_k - y_k)}{2} \tag{16}$$

$$RMSE_x = \sqrt{\sum_{k=1}^{T}\frac{(\hat{x}_k - x_k)^2}{T}}. \tag{17}$$

$$RMSE_y = \sqrt{\sum_{k=1}^{T}\frac{(\hat{y}_k - y_k)^2}{T}}. \tag{18}$$

$$RMSE_{avg} = \frac{(RMSE_x + RMSE_y)}{2} \tag{19}$$

where,

$(\hat{x}_k, \hat{y}_k)$— location estimate for k^{th} time instance,

(x_k, y_k)—real target position at k^{th} time instance.

A sensor node has three important energy-consuming units, namely, a sensor node, processing unit, and RF transceiver [14]. The energy consumption with sensor-based target localization is basically due to the energy consumption from the processing unit ($E_{\mu p}$) and RF transceiver (E_{Transc}). This energy consumption can be expressed as follows [14]:

$$E = E_{Transc} + E_{\mu p} \tag{20}$$

Assuming the total N RSSI measurements with bit duration $^{t}RSSI$ and N ACKs with bit duration ^{t}ACK, the energy consumption for the estimation of the distance for one anchor node is given as follows [14]:

$$E_{anchor} = E_{UP\_Transc} + N * (t_{RSSI} * P_{tx} + t_{ACK} * P_{rx}) + E_{UP\_\mu p} + (t_{RSSI} + t_{ACK}) * P_{on\_\mu p} \tag{21}$$

where,

E_{UP_Transc}—energy needed for waking up a transceiver,
$E_{UP\_\mu p}$—energy needed for waking up a microcontroller,
P_{tx}—transmitter power,
P_{rx}—receiver power,
$P_{on\_\mu p}$—power of the sensor node microcontroller in the active state,
N—Total number of RSSI measurements considered. Here as there are six anchor nodes, so it is 6 for trilateration case, and 3 for SVR case.

An unknown node (here node associated with moving target) expends energy (Eunkn) given by

$$E_{unkn} = E_{UP\_Transc} + N * (t_{ACK} * P_{tx} + t_{RSSI} * P_{rx}) + E_{UP\_\mu p} + (t_{RSSI} + t_{ACK}) * P_{on\_\mu p} \tag{22}$$

For the total energy consumption in RSSI-based localization using multilateration for locating an unknown node (target node here) using six AN's, the total energy consumption is expressed as follows:

$$E_{RSSI} = 6 * (N * E_{anchor} + E_{unkn}) + E_{\mu p\_proc} \tag{23}$$

where $E_{\mu p\_proc}$—energy consumed by the microcontroller in a target node for executing multilateration or the proposed SVR algorithm. From Equations (21)–(23), energy consumption depends on the number of RSSI measurements (N), and is linearly proportional with N.

In this research work, we assumed a 40B length RSSI frame and 11B ACK frame. The energy consumed for a target L&T with trilateration and the proposed SVR models can be calculated using Equations (21) and (22). The typical values of an 802.15.4 compliant RF transceiver were adopted for this analysis, as shown in Table 3. The energy consumed by the microcontroller (associated with a target node) calculating its location was computed from the time duration of the proposed SVR algorithm for processing RSSI measurements from six AN's.

Table 3. Energy-related simulation parameters for the microcontroller and transceiver.

	Microcontroller	
Parameter	**Value**	**Unit**
Current draw in active state	8	mA
Wake up time	1	ms
	Transceiver	
Current draw RX	16	mA
Current draw TX, 3 dB	17	mA
Current draw TX, −17 dB	10	mA
Wake up time	1	ms
Bit Rate	250	kbps

5. Results and Discussion

It is quite logical that the location estimates obtained with any SVR-based model with different kernel functions will have varying performance. Thus, it would be quite interesting to check the impact of different kernel functions on indoor target localization with the proposed SVR-based framework. In this work, we used four kernel functions with the proposed SVR scheme, namely, linear, RBF, polynomial, and sigmoid. These kernel functions were tested separately for the same system setup in case I to case IV. Case I to case IV evaluated the impact of linear, polynomial, RBF, and sigmoid kernel functions on the target localization with the proposed SVR-based model, respectively. The results of case I are provided in Figure 4 and Table 4. In order to differentiate the real target trajectory and location estimations obtained using trilateration, SVR, and SVR+KF, we used "red square", "blue circle", "black plus", and "red plus" markers, respectively, in Figure 4 (case I). The same color markers were used in the rest of the cases considered in this work. To assess the localization accuracy of trilateration and the proposed SVR-based schemes, the RMSE values for x-coordinate estimation and y-coordinate estimation along with average RMSE values and average localization errors were computed for each of the four cases using Equation (16) to Equation (19), respectively. Figure 4a shows the actual target trajectory in the given indoor environment and the estimated trajectories with all the considered localization techniques.

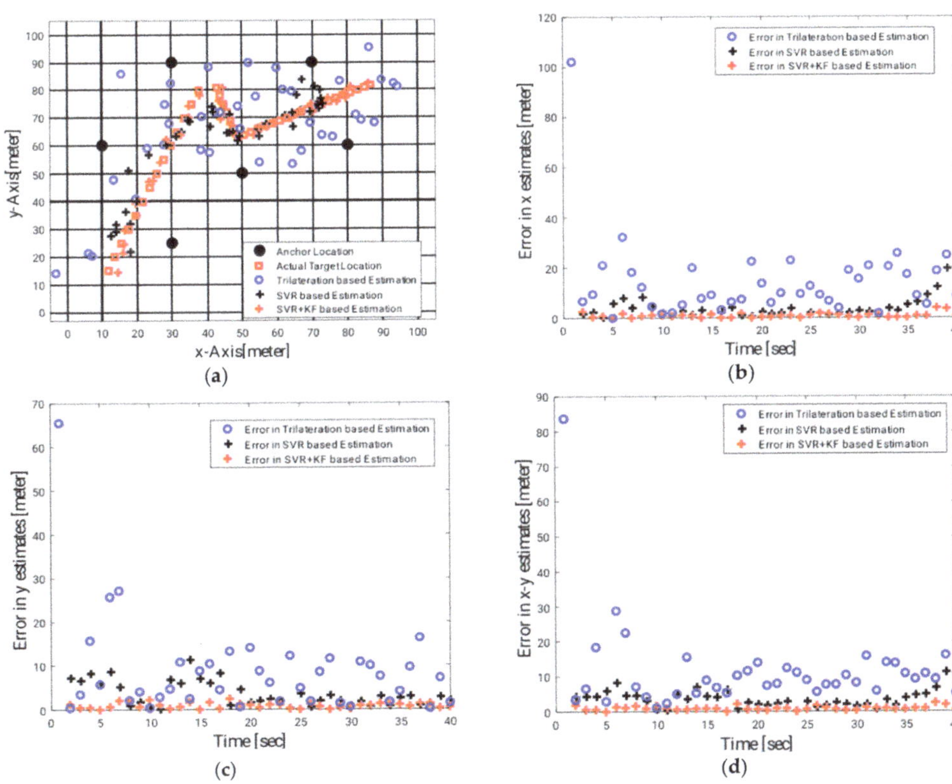

Figure 4. Case I: the impact of linear kernel on SVR-based target localization. (**a**) Location estimates with trilateration, SVR, and SVR+KF against actual target trajectory, (**b**) localization error with trilateration, SVR, and SVR+KF along the x direction, (**c**) localization error with trilateration, SVR, and SVR+KF along the y direction, (**d**) localization error with trilateration, SVR, and SVR+KF along the x–y direction estimates of the mobile target obtained with trilateration, SVR, and SVR+KF.

Table 4. Average localization error and RMSE for case I.

L&T Scheme	$RMSE_x$ (in Meters)	$RMSE_y$ (in Meters)	$RMSE_{avg}$ (in Meters)	*Average Localization Error* (in Meters)	Total Energy Consumption
Trilateration	21.62	14.16	17.89	11.65	2.50 J
SVR (Proposed)	5.95	5.55	5.75	3.92	1.89 mJ
SVR+KF (Proposed)	0.13	0.09	0.11	0.1	2.22 mJ

As discussed in Section 4, the target was assumed to take 40 locations in the given indoor environment during its motion. Figure 4b,c illustrate the localization (location estimation) error for each of these 40 target locations with trilateration, SVR, and SVR+KF for the x-coordinate and y-coordinate for each target, respectively. To obtain the overall 2-D localization performance with trilateration for the proposed SVR-based models, the location estimation errors for the x and y coordinates were averaged to obtain the average localization error. The values of the average localization errors for 40 target positions are shown in Figure 4d. From Figure 4a, it can be seen that the estimations achieved with the trilateration scheme were far away from the corresponding real target positions compared with those of the proposed SVR models. Few location estimates obtained with plain SVR models without KF are close to the corresponding real target positions; however, the remaining estimates are away from the actual target location by 2 to 5 m. Most of the estimates given by the proposed SVR+KF model coincided with the corresponding real target positions. The individual location estimation error for the x-coordinate and y-coordinate corresponding to the actual target locations can be observed in Figures 4c and 4d, respectively. The RMSE values and average position estimation errors were highest, moderate and lowest with trilateration, and the proposed SVR-based schemes, respectively, in the case of linear kernel (case I) (See Table 4). The average RMSE with the SVR-based schemes for the linear kernel function case decreased by 68% and 99%, respectively, against the trilateration-based localization scheme. The average position estimation error with SVR and SVR+KF for the linear kernel function case decreased by approximately 66% and 91%, respectively, as compared with that of the trilateration-based localization scheme. Thus, both SVR-based schemes outperformed the traditional trilateration-based scheme using RSSI measurements.

The case II results with the application sigmoid kernel-based SVR schemes are given in Figure 5 and Table 5. Unlike the case I results, the case II results demonstrated that the target L&T performance with trilateration was superior to the proposed plain SVR-based localization scheme. However, the target-localization performance of the proposed SVR+KF outperformed trilateration by a large margin. The average RMSE with SVR+KF for the sigmoid kernel function decreased by around 98% as compared with that of the trilateration-based localization scheme. The average location estimation error with SVR+KF for the linear kernel function case decreased by around 89% as compared with that of the trilateration-based localization scheme. Although the L&T performance with the proposed SVR+KF outperformed the other considered schemes, the average location estimation error and average RMSE with SVR+KF in case II increased by 48% and 19%, respectively, as compared with that of with SVR+KF in case I. The case III results with the application of the RBF kernel-based SVR schemes are shown in Figure 6 and Table 6. As with the case II results, case III results also showed that the target-localization performance with trilateration was superior to that of the proposed plain SVR-based localization scheme. As in case I and case II, the L&T performance with the proposed SVR+KF outperformed the other considered schemes. However, the average localization error and average RMSE with the SVR+KF-based scheme were high in case III against that of SVR+KF in case I and case II. Thus, at this point, it is clear that the use of the RBF kernel function with the proposed SVR-based schemes is not a good option.

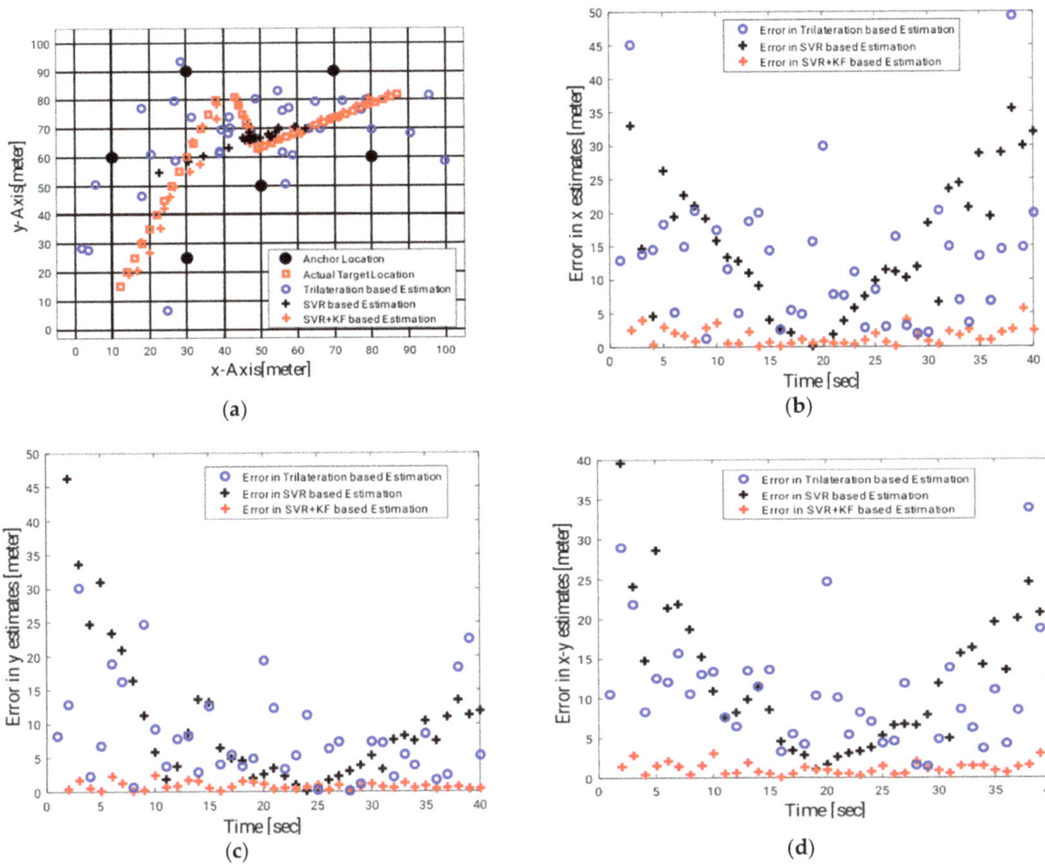

Figure 5. Case II: impact of the sigmoid kernel on SVR-based target localization. (**a**) Location estimates with trilateration, SVR, and SVR+KF against the actual target trajectory, (**b**) localization error with trilateration, SVR, and SVR+KF along the x direction, (**c**) localization error with trilateration, SVR, and SVR+KF along the y direction, (**d**) localization error with trilateration, SVR, and SVR+KF along the x–y direction.

Table 5. Average localization error and RMSE for case II.

L&T Scheme	$RMSE_x$ (in Meters)	$RMSE_y$ (in Meters)	$RMSE_{avg}$ (in Meters)	*Average Localization Error* (in Meters)	Total Energy Consumption
Trilateration	16.61	10.96	13.79	10.70	2.50 J
SVR (Proposed)	18.01	15.71	16.86	12.93	2.31 mJ
SVR+KF (Proposed)	0.39	0.05	0.22	1.22	2.78 mJ

Table 6. Average localization error and RMSE for case III.

L&T Scheme	$RMSE_x$ (in Meters)	$RMSE_y$ (in Meters)	$RMSE_{avg}$ (in Meters)	*Average Localization Error* (in Meters)	Total Energy Consumption
Trilateration	15.03	10.30	12.67	10.15	2.50 J
SVR (Proposed)	18.77	16.63	17.70	13.56	2.56 mJ
SVR+KF (Proposed)	0.64	0.14	0.39	1.52	2.95 mJ

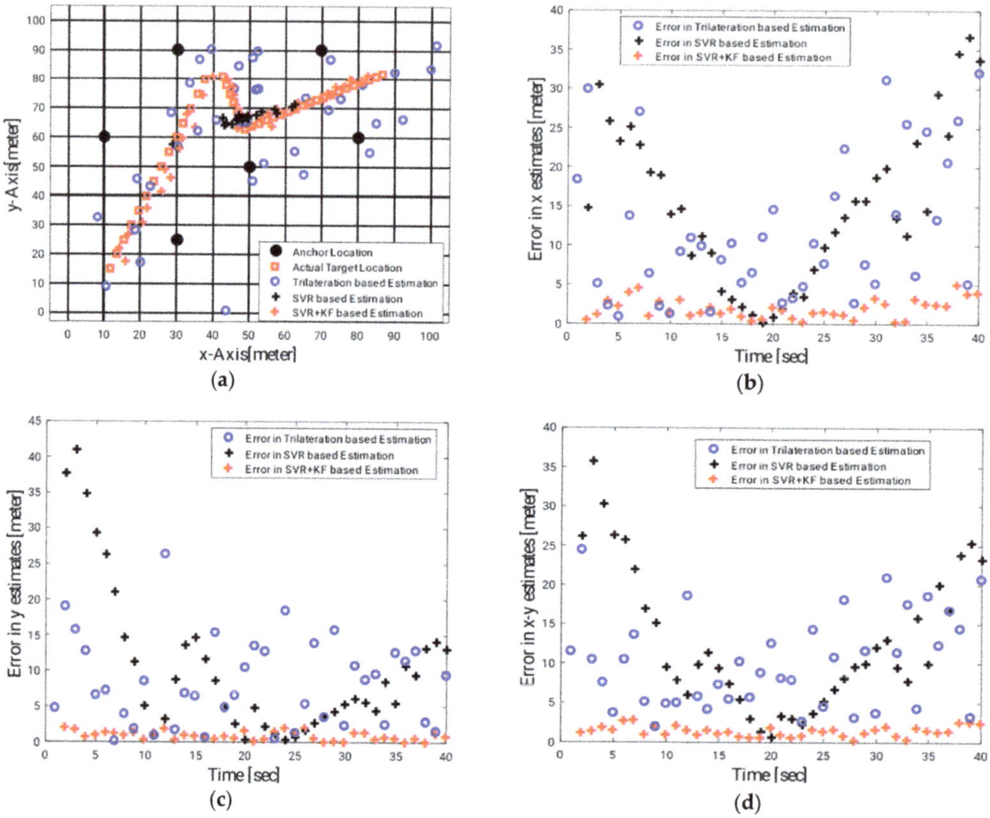

Figure 6. Case III: impact of the RBF Kernel on SVR-based target localization. (**a**) Location estimates with trilateration, SVR, and SVR+KF against the actual target trajectory, (**b**) localization error with trilateration, SVR, and SVR+KF along the x direction, (**c**) localization error with trilateration, SVR, and SVR+KF along the y direction, (**d**) localization error with trilateration, SVR, and SVR+KF along the x–y direction.

The case IV results with the application of polynomial kernel-based SVR schemes are shown in Figure 7 and Table 7. As in case I results, from case IV results it is observed that the target-localization performance with SVR-based schemes was superior to that with trilateration. The RMSE values and average localization errors were highest, moderate and lowest with trilateration, and the proposed SVR-based schemes, respectively, with the polynomial kernel function. The average RMSE with SVR and the SVR+KF for polynomial kernel function case decreased by approximately 36% and 99%, respectively, as compared with that of the trilateration-based localization scheme. The average error in location estimation with SVR and SVR+KF for the polynomial kernel function case decreased by approximately 34% and 91%, respectively, as compared with that of the trilateration-based localization scheme. Thus, both SVR-based schemes outperformed the traditional trilateration-based target-localization approach using RSSI measurements. Comparing the localization performance of the proposed SVR+KF scheme in case I and case IV, it is clearly observed that the target-localization accuracy with the polynomial kernel function-enabled SVR scheme was very high as compared with that of the linear kernel function-enabled SVR scheme. The average location estimation error and average RMSE with SVR+KF in case IV decreased by approximately 5% and 39%, respectively, against that with SVR+KF in case I.

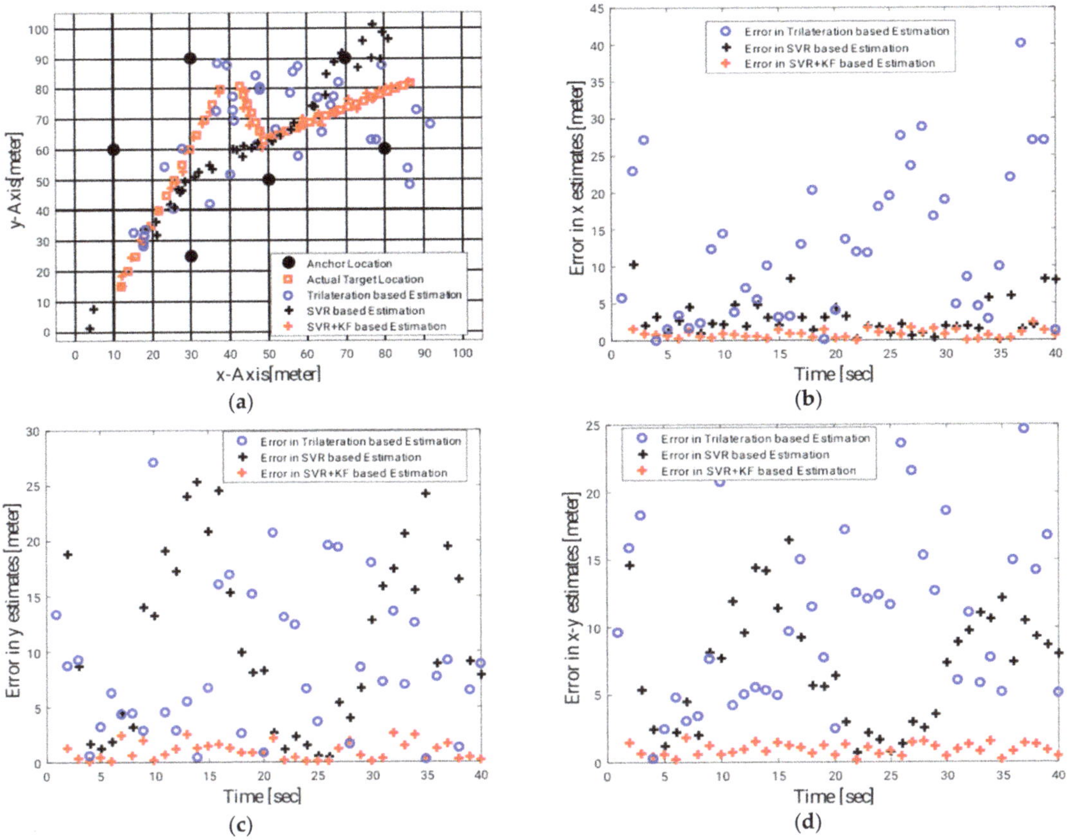

Figure 7. Case IV: impact of polynomial kernel on SVR-based target localization. (**a**) Location estimates with trilateration, SVR, and SVR+KF against the actual target trajectory, (**b**) localization error with trilateration, SVR, and SVR+KF along the x direction, (**c**) localization error with trilateration, SVR, and SVR+KF along the y direction, (**d**) localization error with trilateration, SVR, and SVR+KF along the x–y direction.

Table 7. Average localization error and RMSE for case IV.

L&T Scheme	$RMSE_x$ (in Meters)	$RMSE_y$ (in Meters)	$RMSE_{avg}$ (in Meters)	*Average Localization Error* (in Meters)	Total Energy Consumption
Trilateration	16.07	10.94	13.50	10.71	2.50 J
SVR (Proposed)	3.95	13.45	8.70	7.05	1.68 mJ
SVR+KF (Proposed)	0.11	0.03	0.07	0.95	2.12 mJ

Thus, after discussing the target-localization results in case I to case IV, it is confirmed that the selection of kernel function in the SVR-based target-localization model had a significant impact on target tracking accuracy. In this research work, a number of important research findings can be noted. Regarding the proposed SVR+KF scheme, the highest localization accuracy could be seen with the polynomial kernel function (case IV) as compared with that of the rest of the other kernel functions considered in this study. Whereas, regarding the proposed plain SVR scheme, the highest localization accuracy could be seen with the linear kernel function (case I) as compared with that of the rest of the other considered kernel functions. The time complexity of trilateration, SVR, and SVR+KF models were found to be 4 milliseconds, 2.9 milliseconds, and 4.2 milliseconds, respectively, by using the tic-toc command of MATLAB. Thus, proposed model does not add more complexity as compared to trilateration. As discussed earlier, in moderate and high localization accuracy-demanding applications, the proposed SVR-based target L&T models can be selected. The proposed SVR-based schemes utilize RSS measurements, which are very fluctuating in nature. Therefore, for each trial in case I to case IV, different localization results were obtained. Thus, in order to avoid misleading conclusions, the results provided in this research article are based on an average of 50 trials of each case. We believe that for different indoor environmental setups, the localization results can be different. However, the research findings in this paper with the proposed SVR-based L&T schemes can be a very good guide to build specific SVR-based models to solve the problem of target localization and tracking for any given indoor setup.

6. Conclusions

This paper provided an SVR-based target-localization scheme which can deal with noise uncertainty in RSSI measurements and high-velocity variation in target motion. Rigorous simulations were conducted to test the impact of the kernel function with the proposed SVR-based schemes on indoor localization performance. We tested four popular kernel functions of SVM, namely, linear, sigmoid, RBF, and polynomial. The simulation results proved that for the linear and polynomial kernel function the proposed SVR-based target-localization model demonstrates superior localization performance along with less energy consumption involved in localization as compared with that of trilateration. Generally, based on demand of the underlying application, the need for accuracy in localization performance may vary. For the applications in which a localization accuracy of around 5 to 8 m is needed, the proposed plain SVR-based scheme is a good lightweight option for indoor target localization. The proposed SVR+KF target L&T scheme is more suitable for applications which demand centimeter-level target-localization accuracy in a given indoor environment. The proposed research work can be extended in many ways. The proposed SVR models may be applied to solve the multi-target tracking (MTT) problem for indoor environments. We believe that as the target trajectory changes in a considered indoor environment, the localization accuracy may vary. Even changes in the WSN operating area will have an impact on the localization accuracy. In such scenarios, we believe that by varying number of AN's, and training the proposed model for a given setup again, the model can estimate the target track satisfactorily.

Author Contributions: Conceptualization, J.P.M.; supervision, D.D. and S.B.G.; original draft, review and editing, S.R.J. and S.S.A.; validation, A.S.R. and S.B.G.; review and editing, A.S.R., S.B.G., M.S.R., T.C.M., C.V., G.S. All authors have read and agreed to the published version of the manuscript.

Funding: This research work has not received any external funding.

Data Availability Statement: The authors declare that we can provide all the necessary data upon request.

Acknowledgments: This paper was partially supported by UEFISCDI Romania and MCI through BEIA projects AutoDecS, SOLID-B5G, T4ME2, DISAVIT, PIMEO-AI, AISTOR, MULTI-AI, ADRIATIC, Hydro3D, PREVENTION, DAFCC, EREMI, ADCATER, MUSEION, FinSESCo, iPREMAS, IPSUS, U-GARDEN, CREATE and by European Union's Horizon 2020 research and innovation program under grant agreements No. 883522 (S4ALLCITIES) and No. 101016567 (VITAL-5G). The results were obtained with the support of the Ministry of Investments and European Projects through the Human Capital Sectoral Operational Program 2014-2020, Contract no. 62461/03.06.2022, SMIS code 153735. This work is supported by Ministry of Research, Innovation, Digitization from Romania by the National Plan of R & D, Project PN 19 11, Subprogram 1.1. Institutional performance-Projects to finance excellence in RDI, Contract No. 19PFE/30.12.2021 and a grant of the National Center for Hydrogen and Fuel Cells (CNHPC)—Installations and Special Objectives of National Interest (IOSIN).

Conflicts of Interest: The authors declare no conflict of interest.

References

1. Huang, H.; Gartner, G.; Krisp, J.M.; Raubal, M.; Van De Weghe, N. Location based services: Ongoing evolution and research agenda. *J. Locat. Based Serv.* **2018**, *12*, 63–93. [CrossRef]
2. Jondhale, S.R.; Jondhale, A.S.; Deshpande, P.S.; Lloret, J. Improved trilateration for indoor localization: Neural network and centroid-based approach. *Int. J. Distrib. Sens. Netw.* **2021**, *17*, 15501477211053997. [CrossRef]
3. Jondhale, S.R.; Wakchaure, M.A.; Agarkar, B.S.; Tambe, S.B. Improved generalized regression neural network for target localization. *Wirel. Pers. Commun.* **2022**, *125*, 1677–1693. [CrossRef]
4. Akyildiz, I.; Su, W.; Sankarasubramaniam, Y.; Cayirci, E. A survey on sensor networks. *IEEE Commun. Mag.* **2002**, *40*, 102–114. [CrossRef]
5. Jondhale, S.R.; Sharma, M.; Maheswar, R.; Shubair, R.; Shelke, A. Comparison of Neural Network Training Functions for RSSI based Indoor Localization Problem in WSN. In *Handbook of Wireless Sensor Networks: Issues and Challenges in Current Scenario's, Part of AISC*; Springer Nature: Berlin, Germany, 2020; Volume 1132, pp. 112–133.
6. Jondhale, S.R.; Maheswar, R.; Lloret, J. Survey of Existing RSSI-Based L&T systems. In *EAI/Springer Innovations in Communication and Computing*; Springer: Cham, Switzerland, 2022; pp. 49–64.
7. Jondhale, S.R.; Maheswar, R.; Lloret, J. *Received Signal Strength Based Target Localization and Tracking Using Wireless Sensor Network*; EAI/Springer Innovations in Communication and Computing Series; Springer International Publishing: Midtown Manhattan, NY, USA, 2022.
8. Yan, X.; Luo, Q.; Yang, Y.; Liu, S.; Li, H.; Hu, C. ITL-MEPOSA: Improved trilateration localization with minimum uncertainty propagation and optimized selection of anchor nodes for wireless sensor networks. *IEEE Access* **2019**, *7*, 53136–53146. [CrossRef]
9. Jondhale, S.R.; Deshpande, R.S. Tracking target with constant acceleration motion using kalman filtering. In Proceedings of the International Conference on Advances in Communication and Computing Technology (ICACCT), Sangamner, India, 8–9 February 2018; pp. 581–587.
10. Nessa, A.; Adhikari, B.; Hussain, F.; Fernando, X.N. A survey of machine learning for indoor positioning. *IEEE Access* **2020**, *8*, 214945–214965. [CrossRef]
11. Chriki, A.; Touati, H.; Snoussi, H. SVM-based indoor localization in Wireless Sensor Networks. In Proceedings of the 2017 13th International Wireless Communications and Mobile Computing Conference (IWCMC), Valencia, Spain, 26–30 June 2017; pp. 1144–1149. [CrossRef]
12. Jondhale, S.R.; Mohan, V.; Sharma, B.B.; Lloret, J.; Athawale, S.V. Support vector regression for mobile target localization in indoor environments. *Sensors* **2022**, *22*, 358. [CrossRef] [PubMed]
13. Zhang, L.; Kuang, Z.; Wang, Z.; Yang, Z.; Zhang, S. A node three-dimensional localization algorithm based on RSSI and LSSVR parameters optimization. *Syst. Sci. Control Eng.* **2020**, *8*, 477–487. [CrossRef]
14. Moravek, P.; Komosny, D.; Simek, M.; Girbau, D.; Lazaro, A. Energy analysis of received signal strength localization in wireless sensor networks. *Radioengineering* **2011**, *20*, 937–945.
15. Meghani, S.K.; Asif, M.; Awin, F.; Tepe, K. Empirical based ranging error mitigation in IR-UWB: A fuzzy approach. *IEEE Access* **2019**, *7*, 33686–33697. [CrossRef]
16. Pascacio, P.; Casteleyn, S.; Torres-Sospedra, J.; Lohan, E.S.; Nurmi, J. Collaborative indoor positioning systems: A systematic review. *Sensors* **2021**, *21*, 1002. [CrossRef] [PubMed]
17. Liu, Z.; Dai, B.; Wan, X.; Li, X. Hybrid wireless fingerprint indoor localization method based on a convolutional neural network. *Sensors* **2019**, *19*, 4597. [CrossRef] [PubMed]
18. Wang, H.; Li, J.; Cui, W.; Lu, X.; Zhang, Z.; Sheng, C.; Liu, Q. Mobile robot indoor positioning system based on K-ELM. *J. Sens.* **2019**, *2019*, 7547648. [CrossRef]
19. Zhou, C.; Wieser, A. Application of backpropagation neural networks to both stages of fingerprinting based WIPS. In Proceedings of the 2016 Fourth International Conference on Ubiquitous Positioning, Indoor Navigation and Location Based Services (UPINLBS), Shanghai, China, 2–4 November 2016; pp. 207–217. [CrossRef]
20. Nguyen, X.; Jordan, M.; Sinopoli, B. A kernel-based learning approach to ad hoc sensor network localization. *ACM Trans. Sens. Networks* **2005**, *1*, 134–152. [CrossRef]
21. Tran, D.A.; Nguyen, T. Localization in wireless sensor networks based on support vector machines. *IEEE Trans. Parallel Distrib. Syst.* **2008**, *19*, 981–994. [CrossRef]
22. Ma, C.; Yang, M.; Jin, Y.; Wu, K.; Yan, J. A new indoor localization algorithm using received signal strength indicator measurements and statistical feature of the channel state information. In Proceedings of the 2019 International Conference on Computer, Information and Telecommunication Systems (CITS), Beijing, China, 28–31 August 2019. [CrossRef]
23. Yoo, J.; Kim, H.J. Target localization in wireless sensor networks using online semi-supervised support vector regression. *Sensors* **2015**, *15*, 12539–12559. [CrossRef] [PubMed]
24. Stella, M.; Russo, M.; Begusic, D. Location determination in indoor environment based on RSS fingerprinting and artificial neural network. In Proceedings of the 2007 9th International Conference on Telecommunications, Zagreb, Croatia, 13–15 June 2007. [CrossRef]
25. Jondhale, S.R.; Deshpande, R.S. Modified Kalman filtering framework based real time target tracking against environmental dynamicity in wireless sensor networks. *Ad-Hoc Sens. Wirel. Netw.* **2018**, *40*, 119–143.

26. Jondhale, S.R.; Deshpande, R.S. GRNN and KF framework based real time target tracking using PSOC BLE and smartphone. *Ad Hoc Netw.* **2019**, *84*, 19–28. [CrossRef]
27. Jondhale, S.R.; Deshpande, R.S. Kalman filtering framework-based real time target tracking in wireless sensor networks using generalized regression neural networks. *IEEE Sensors J.* **2018**, *19*, 224–233. [CrossRef]
28. Sampathkumar, A.; Mulerikkal, J.; Sivaram, M. Glowworm swarm optimization for effectual load balancing and routing strategies in wireless sensor networks. *Wirel. Netw.* **2020**, *26*, 4227–4238. [CrossRef]

energies

Article

Evaluation of Particulate Matter (PM) Emissions from Combustion of Selected Types of Rapeseed Biofuels

Joanna Szyszlak-Bargłowicz [1], Jacek Wasilewski [1], Grzegorz Zając [1,*], Andrzej Kuranc [1], Adam Koniuszy [2] and Małgorzata Hawrot-Paw [2]

1 Department of Power Engineering and Transportation, Faculty of Production Engineering, University of Life Sciences in Lublin, Głęboka 28, 20-612 Lublin, Poland
2 Department of Renewable Energy Sources Engineering, West Pomeranian University of Technology in Szczecin, Papieża Pawła VI 1, 71-459 Szczecin, Poland
* Correspondence: grzegorz.zajac@up.lublin.pl

Abstract: The manuscript describes the results of an experimental study of the level of PM (particulate matter) emissions arising from the combustion of two selected types of biomass (i.e., rapeseed straw pellets and engine biofuel (biodiesel, FAME)), which were derived from rapeseed. The PM emissions from the combustion of biofuels were compared with those obtained from the combustion of their traditional counterparts (i.e., wood pellets and diesel fuel). Both types of pellets were burned in a 10 kW boiler designed to burn these types of fuels. The engine fuels tested were burned in a John Deere 4045TF285JD engine mounted on a dynamometer bench in an engine dyno, under various speed and load conditions. A Testo 380 analyzer was used to measure the PM emission levels in boiler tests, while an MPM4 particle emission meter was used in the engine tests. The combustion (under rated conditions) of rapeseed straw pellets resulted in a significant increase in PM emissions compared to the combustion of wood pellets. The PM emissions during the combustion of wood pellets were 15.45 $mg \cdot kg^{-1}$, during the combustion of rapeseed straw pellets, they were 336 $mg \cdot kg^{-1}$, and the calculated emission factors were 44.5 $mg \cdot MJ^{-1}$ and 1589 $mg \cdot MJ^{-1}$, respectively. In the engine tests, however, significantly lower particulate emissions were obtained for the evaluated biofuel compared to its conventional counterpart. The combustion of rapeseed oil methyl esters resulted in a 40–60% reduction in PM content in the exhaust gas on average for the realized engine speeds over the full load range compared to the combustion of diesel fuel.

Keywords: particulate matter (PM); rapeseed straw pellets; biodiesel; combustion; PM emission levels; pellet boiler; diesel engine; engine dynamometer

Citation: Szyszlak-Bargłowicz, J.; Wasilewski, J.; Zając, G.; Kuranc, A.; Koniuszy, A.; Hawrot-Paw, M. Evaluation of Particulate Matter (PM) Emissions from Combustion of Selected Types of Rapeseed Biofuels. *Energies* **2023**, *16*, 239. https://doi.org/10.3390/en16010239

Academic Editors: Vadim Bolshev, Vladimir Panchenko, Nallapaneni Manoj Kumar, Pandian Vasant, Igor S. Litvinchev and Prasun Chakrabarti

Received: 30 November 2022
Revised: 20 December 2022
Accepted: 22 December 2022
Published: 26 December 2022

1. Introduction

A closed-loop economy is related to the need to maintain the high value and quality of resources, products, and materials for as long as possible, and to minimize the amount of waste produced by managing it efficiently. A common element in all activities is the effort to close the circulation of materials in the economy [1]. A significant direction of the use of agricultural raw materials in addition to the production of liquid biofuels or the biocomponents of motor fuels is their use in the energy sector. An important role is played by solid biofuels, which are mostly residues from the agricultural and forestry industries, which are components of solid energy fuels or are fuels in their own right. Problems related to these types of energy raw materials not only include their qualitative assessment, but also the impact of their use on the environment [2,3].

Rapeseed straw, which is a waste product in the production of rapeseed oil and biodiesel, can be a valuable energy feedstock with a high calorific value [4–6]. Undeveloped rapeseed straw can also be used in the pressure agglomeration of biomass during the production of compact solid biofuels [7]. Rapeseed straw pellets can be an alternative to wood pellets used in low-power automatically supplied boilers. Such activities are in line with the main objectives of the EU Common Agricultural Policy, in relation to practices that promote environmental and climate protection [8].

In urban conditions, biomass combustion is a major source of particulate matter during the winter seasons, accounting for up to 30% of PM10 [7]. Particulate matter contributes significantly to the total emissions of atmospheric air pollutants accompanying fuel combustion in low-power boilers. The results of studies of the fractional composition of dust emitted from domestic furnaces indicate a significant contribution, depending on the type of furnace and its mode of use, of fine fractions of particulate matter including PM10. Emissions from biomass combustion in domestic furnaces involve an unaccounted-for share of TSP (total suspended particulate) dust emissions [9]. Small domestic boilers and stoves remain the main emitters of PM in many countries, especially during the winter season. In the EU, more than 40% of the energy produced from solid biomass comes from combustion in domestic boilers and stoves [9]. Depending on the local fuel availability, domestic boilers are fueled by wood, biomass pellets, or coal of various types and qualities. During the heating season, in some EU countries, the share of total PM (particulate matter) emissions emitted from biomass combustion in domestic appliances can exceed 80% [10]. Residential biomass combustion is one of the largest sources of fine particles in the global troposphere, which has serious impacts on the air quality, climate, and human health. Because of this correlation and the increasing use of biomass for energy purposes through combustion, emissions from biomass combustion, particularly PM, require further study [8]. It is difficult to obtain quantitative estimates of the contribution of this source to airborne particulate levels because the emission factors vary widely depending on the type of wood, combustion equipment, and operating conditions [10]. There are two main sources of particulate matter emitted from domestic and small district heating boilers: (1) particles formed by incomplete combustion (soot, condensable organic particles—tars) and char; and (2) particles from the inorganic material in the fuel-ash [11].

The analysis of literature reports on the mechanisms of the formation of solid particles in the process of biomass combustion proves that PM emissions depend on the combustion technology, especially on the physical and chemical properties of the biomass being burned [9]. Solid particles are formed mainly from evaporating inorganic components such as KCl, which in the presence of SO_2 undergo further chemical transformations (e.g., to K_2SO_4). The composition of the emitted inorganic submicron solid particles mainly includes potassium, chlorine, sulfur, and oxygen, and the process of releasing alkali metals during biomass combustion depends on the mutual ratio of the content of chlorine and alkali metals as well as the presence of sulfur and nitrogen in the fuel [11,12].

Another way to use biomass for energy purposes is the use of biofuels in transportation. Biodiesel (FAME) is produced by a transesterification reaction with methanol, resulting in a mixture of fatty acid methyl esters [13,14].

Particulate matter is also formed during the operation of internal combustion engines, mainly from diesel engines, which can be fueled with biodiesel or fuel containing biodiesel in the form of a biocomponent.

They contain the following components (Figure 1) [15,16]:

- *Insoluble Organic Fraction* (IOF—a part of INSOL), in other words, carbon in the form of soot and products of incomplete combustion of fuel and oil additives;
- *Insoluble inorganic fraction* (INSINOF—a part of INSOL), which consists of ashes, sulfates, trace elements such as iron, phosphorus, calcium, chromium, etc., and mechanical impurities from the environment;
- *Soluble Organic Fraction* (SOF), organic substances absorbed on soot particles (mainly hydrocarbons formed from the incomplete combustion of fuel PM_{FUEL} and oil PM_{LUBE});
- *Soluble Inorganic Fraction* (SINOF), resulting mainly from the presence of sulfur in the fuel, from which sulfuric acid is formed following combustion and because of the presence of water vapor;

- *Soluble Organic Fraction* (SOF), which consists mainly of unburned hydrocarbons resulting from local oxygen deficiency or excess from flame extinction near the cooler cylinder walls or from a drop in charge temperature during expansion.

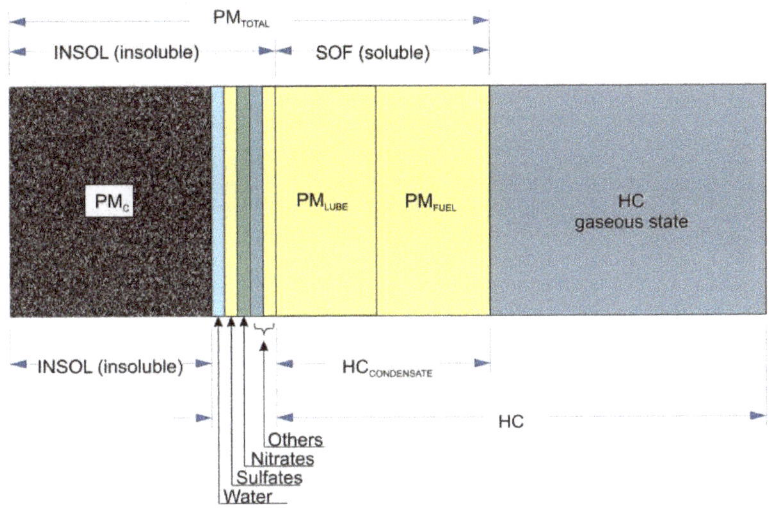

Figure 1. Diagram of the structure of a solid particle.

Typically, diesel engine exhaust contains small (10–80 μm), single elementary particles of soot in the shape of a sphere called nanoparticles and large (10–50 μm) forming clusters of these particles in the form of soot agglomerates or aggregates (more than 100 μm) [15].

The consequence of air pollution is a reduced life expectancy. Particulate emissions are one of the main causes of smog. Particulate matter (PM) air pollution leads to premature deaths from heart disease, stroke, and cancer, and causes acute respiratory infections [17–19]. Air pollution is estimated to cause seven million deaths worldwide each year [20]. In Europe, air pollution is the biggest environmental threat to human health, a leading cause of premature births, deaths, and many diseases [21,22]. Particulate matter PM10 and PM2.5 (particulate matter less than 10 μm and 2.5 μm in size, respectively) are considered especially hazardous to human health, causing respiratory diseases (such as asthma) and even death [23].

The literature data on the emissions and mechanisms of particulate matter from various biofuels are quite scattered, and further work in this area is needed to study, describe, and organize them. In the absence of reports comprehensively treating the effect of using solid and liquid rapeseed biofuels on particulate emissions, it was decided to study them and compare them with fuels considered conventional. The purpose of the study was to analyze the level of PM emissions produced when burning, in different power equipment and varying operating conditions, different types of biofuels made from rapeseed. The biofuels studied were biodiesel (FAME—rapeseed oil methyl esters) used to power compression-ignition engines, and pellets made from rapeseed straw, which is a by-product of edible oil production that can be used for energy purposes. In addition, the obtained test results were compared with those of particulate emissions during the combustion of conventional diesel fuel and wood pellets as the primary fuel for an automatic low-power pellet boiler. This comprehensive approach to the energy use of biofuels obtained on the basis of rapeseed in comparison with their conventional counterparts is a continuation of emissivity studies, the results of which in relation to selected greenhouse gases were presented in [6].

2. Materials and Methods

2.1. Boiler Tests

Two types of solid biofuels were used in the boiler emission tests. Rapeseed straw pellets were produced for the study, and commercial wood pellets of ENplus A1 quality grade were obtained from coniferous trees. In both cases, the pellets were 6 mm in diameter. To characterize the pellets, proximate and ultimate analyses were performed and LHV and HHV were determined. The results are shown in Table 1.

Table 1. Chosen characteristics of the tested pellets.

Parameter	Wood Pellets	Rape Straw Pellets
Carbon (%)	49.5	40.1
Hydrogen (%)	6.06	5.8
Nitrogen (%)	0.17	0.8
Sulfur (%)	0.02	0.31
Oxygen (%) *	38.25	33.19
Moisture (%)	5.7	9.4
Volatile matter (%)	84.45	64.7
Ash (%)	0.3	10.4
LHV ($MJ \cdot kg^{-1}$)	17.89	14.76
HHV ($MJ \cdot kg^{-1}$)	19.95	15.97

The tests were carried out with the use of a 10 kW automatically fed boiler by Greń (Pszczyna, Poland). Primary and secondary air was fed by a fan controlled by the boiler controller.

The exhaust gas flow rate was determined by the measurement of the gas velocity with an L-type Pitot tube (Testo SE & Co. KGaA, Titisee-Neustadt, Germany). Pellet consumption was measured during combustion using a scale placed under the stove platform. The test stand is shown in Figure 2 (a schematic is available in [6]).

After the boiler was started, the fuel dosage was gradually increased to achieve the rated operating parameters and a stable flue gas temperature, which was the criterion for achieving proper operating conditions. The boiler was then operated for 1 h at full load, during which the particulate emissions were monitored.

The TSP mass was measured using a particulate matter measurement system Testo 380 combined with Testo 330-2 LL (Testo SE & Co. KGaA, Germany). The analyzer's probe was mounted in the chimney on a straight section of the flue. The exhaust gas analyzer performed automatic measurements at a frequency of 10 s. In order to compare emissions, the results obtained in $mgNm^{-3}$ were converted to emission factors related to a unit of fuel mass (1) and a unit of energy (2):

$$EF_{PM} = \frac{C \cdot V_{Total}}{m_{fuel}}, \left(\mathrm{mg_{PM} \cdot kg_{fuel}^{-1}}\right), \tag{1}$$

$$EF_{PM} = \frac{C \cdot V_{Total}}{m_{fuel} \cdot HHV}, \left(\mathrm{mg_{PM} \cdot MJ^{-1}}\right), \tag{2}$$

where

C is the average concentration of PM ($mg \cdot m^{-3}$);
V_{Total} is the total volume of the gas sampled during the experiment (m^3);
m_{fuel} is the mass of dry fuel consumed (kg);
HHV is the higher heating value ($MJ \cdot kg^{-1}$).

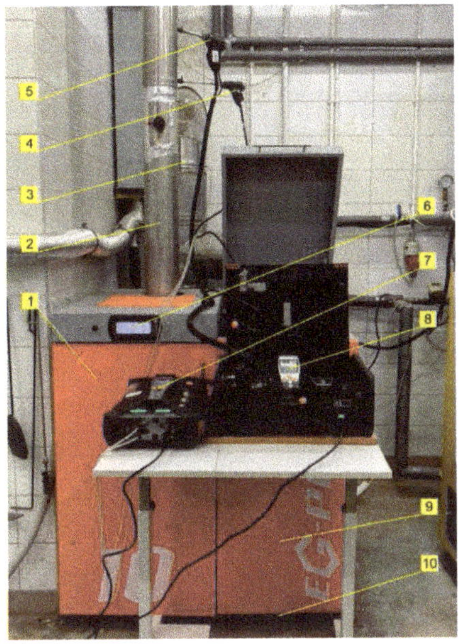

Figure 2. Boiler test stand: 1—test boiler, 2—chimney, 3—Pitot tube, 4—fume probe, 5—flow probe, 6—boiler controller, 7—Testo 350—flow meter, 8—Testo 380—PM meter, 9—pellet reservoir, 10—scales.

2.2. Engine Tests

Rapeseed biofuel FAME (fatty acid methyl esters)—B100 and Efecta Diesel Fuel—DF, were used in the study. Table 2 contains the selected physical and chemical properties of both fuels.

Table 2. Chosen characteristics of the tested fuels.

Parameter	DF	B100
Cetane number	51.4	52.1
Density @ 15 °C (kg·m^{-3})	835	883
Viscosity @ 40 °C (mm^2·s^{-1})	2.6	4.47
Flash point (°C)	69	120
FAME content (% w/w)	6.8	98.8
Carbon (%)	85.7	76.9
Hydrogen (%)	10.6	11.9
Oxygen (%)	2.4	10.3
LHV (MJ·kg^{-1})	43.51	37.92
HHV (MJ·kg^{-1})	45.84	40.36

The tests were carried out on a John Deere 4045TF285JD internal combustion engine mounted on a test stand equipped with an eddy current brake (Figure 3). A schematic of the stand is available in [6].

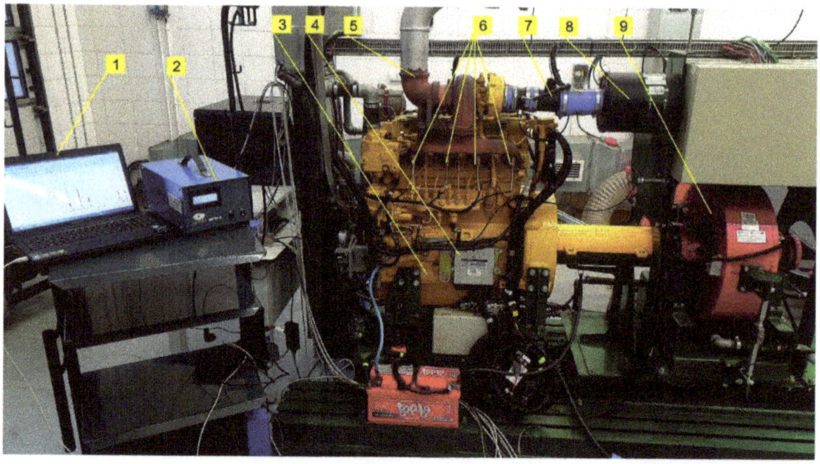

Figure 3. Engine dynamometer test stand: 1—PC-data recorder, 2—MPM4, 3—John Deere engine, 4—engine management unit, 5—exhaust fumes outport, 6—fumes temperature probes, 7—air flow meter, 8—air inlet filter, 9—eddy current engine brake.

An electromotor brake of the AMX—200/6000 type was mounted on the dynamometer bench, with a maximum absorbed power of 200 kW. A strain gauge actuator was mounted on an arm attached to the brake body. Engine speed was measured using an inductive sensor. The test stand was equipped with an ATMX2400 type gravimetric fuel gauge with fuel conditioning. It was also equipped with a system for indexing the engine and recording the rapidly varying pressures of the working medium in the cylinder. The dynamometer's control room, in addition to controlling the operation of the engine-brake unit, allowed for continuous recording of the measured parameters, their visualization, and storage in computer memory. The basic specifications of the tested engine are presented in Table 3.

Table 3. Chosen parameters of the John Deere 4045TF285JD engine.

Parameter	Characteristics
Engine type	Self-ignition engine
Engine displacement	4.5 dm^3
Injection system	direct injection
Fuel System	Mechanical governor
Aspiration	Turbocharged
Cylinder arrangement and number	In-Line, 4-Cycle
Compression ratio	19:1
Nominal power	74 kW
Nominal speed	2400 rpm
Peak torque	353 Nm
Peak torque speed	1600 rpm

Measurements of the PM content in the exhaust gases were carried out based on the engine load characteristics over the full load range, with speeds varying from 1400 to 2400 rpm, in 250 rpm increments. The study used the MPM4 particulate emission meter from MAHA, dedicated to continuous measurement of particulate matter emissions in the exhaust of automotive engines, particularly diesel engines.

The MPM4 meter uses a method of detecting monochromatic radiation (LLSP—laser light scattering photometry) reflected from particles in the exhaust gas, which is an alternative to gravimetric methods (Figure 4).

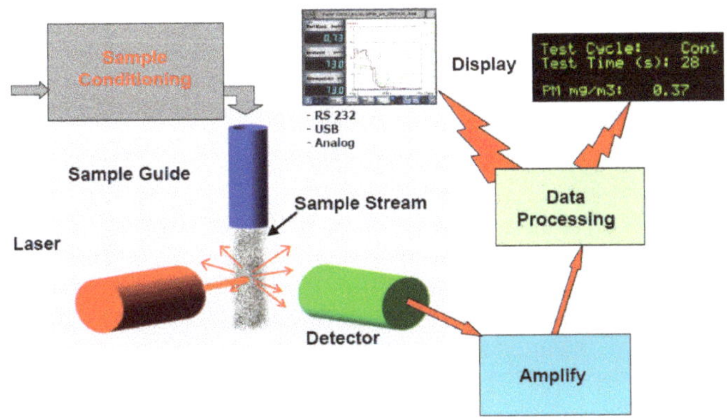

Figure 4. The idea of LLSP—laser light scattering photometry—method used in MPM4 (MPM4_New.pdf).

The measurement results are expressed in mg·m^{-3}. The measurement range was from 0 to 700 mg·m^{-3} and allowed for testing of both modern engines equipped with a diesel particulate filter (DPF) as well as older units and research engines without additional exhaust treatment [24].

2.3. Statistical Analysis

The obtained results of the boiler and engine tests were subjected to statistical analysis using the Statistica ver. 13 program (TIBCO Software Inc., Palo Alto, CA, USA, 2017). The influence of the type of fuel on the PM emission was analyzed using analysis of variance (ANOVA) at a significance level of $\alpha = 0.05$.

3. Results and Discussion

3.1. Boiler Test Results

During the boiler test combustion of the tested biofuels, significantly higher PM emissions were found for the combustion of rapeseed straw pellets (Figure 5).

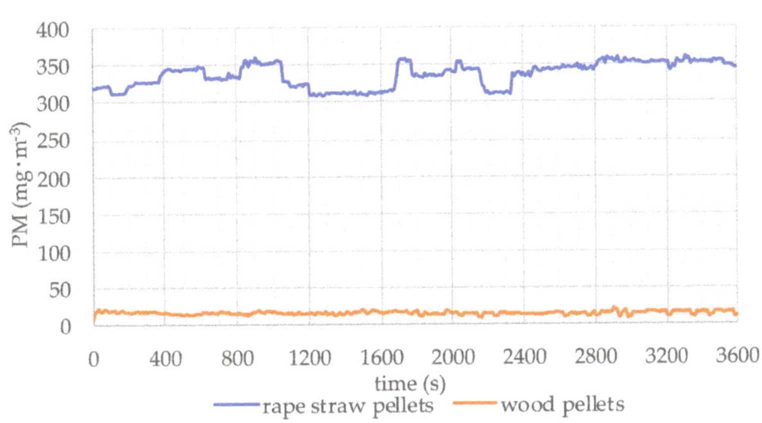

Figure 5. PM emissions during the tests of wood pellets and rapeseed straw pellets.

The average particulate matter emission during the tests of wood pellets was 15.45 mg·m^{-3}, while the average emission during the tests of rapeseed straw pellets was an order of magnitude higher, at 336.9 mg·m^{-3}. Thus, the values of PM emission during the tests of rapeseed straw pellets exceeded many times the emission values that occurred during the tests of wood pellets alone. This was confirmed by similar studies conducted by Chojnacki et al. [25], who found particulate emissions when burning pine pellets of 22.5 mg·m^{-3}, and 218.9 mg·m^{-3} when burning rapeseed straw petals. As reported by Young et al. [26], unlike wood, agricultural solid biofuel had a higher ash content (especially higher alkali metal content), which led to higher PM emissions. In their research, Carroll and Finnan [27] found that the total TSP emissions obtained during wood combustion were 22–51 mg·m^{-3}, while for rapeseed straw biomass, the value rose to 311–399 mg·m^{-3}; they also found similarly high TSP emission values for wheat straw of 253–281 mg·m^{-3}, and barley straw of 251–280 mg·m^{-3}. Garcia-Maraver et al. [28] found that the total particulate emissions ranged from 50 to 100 mg·m^{-3} when pine pellets were burned, while that value increased to 100–600 mg·m^{-3} when olive biomass waste pellets were burned. This was also confirmed in the work of [29], where higher TSP and PM10 emissions were related to agricultural and horticultural biomass combustion, and in the work of [30], where emissions during wood pellet combustion were 104–143 mg·m^{-3}, and during coffee ground pellet combustion, they were 1071–1472 mg·m^{-3}.

Differences in the chemical composition and ash content in the tested pellets resulted in significantly higher PM emissions during the combustion of rapeseed straw pellets.

As noted by [31], there is a significant variation in the chemical composition of biofuels made of different types of biomass. The content of alkali metals, chlorine, and sulfur is higher in cultivated plants than in woody biomass, which results from the cultivation conditions (fertilization).

On the other hand, wood biomass may contain a higher content of heavy metals, which is due to the long vegetation period of trees and the lower pH of forest soils, which increases the solubility of most heavy metal salts.

The tested pellets were also characterized by different humidity, although these differences were not significant. Rapeseed straw pellets (9.4%) had a higher humidity than the wood pellets (5.7%). However, in both cases, the humidity was at a level enabling the direct combustion of these biofuels (<20%) and below the maximum humidity ($\leq$10) specified in the standard ISO 17225-1:2021-11 for wood pellets.

In addition, the combustion of agrobiomass in grate furnaces encounters difficulties, because usually at a temperature of approx. 800 °C, slag is formed [32,33]. This phenomenon comes from the chemical composition of biomass and ash and makes the furnace operating difficult [34–36].

The best agrobiomass combustion technique is two-stage combustion (gasification and process gas combustion) [37]. However, the market lacks boilers with two-stage combustion with a capacity of up to 50 kW, designed to combust pellets of agrobiomass, which is why boilers with grate furnaces dedicated to wood pellets were used. However, in such situations, the combustion of agrobiomass can be problematic, because the temperature in the furnace often exceeds the sintering temperature of the agrobiomass ash and the slag is formed. Thus, the combustion temperature affects the formation of slag, which can also affect the level of PM emissions from the combustion of this type of fuel.

In order to compare and interpret the obtained test results and compare them with the literature reports, the EF emission factors were calculated related to the mass of fuel burned (mg·kg^{-1}) and energy obtained (mg·MJ^{-1}) (Table 4). Emission factors are a relative measure and can be used to assess emissions from various sources of air pollution. Their knowledge is important in developing pollution control strategies and assessing the practicability of burning a specific fuel.

Table 4. Emission factors determined for the tested pellets.

Specification	Wood Pellets	Rape Straw Pellets
PM (mg·m^{-3})	15.45	336.9
EF (mg·kg^{-1})	797	23,282
EF (mg·MJ^{-1})	44.5	1589

In the study in question, the rates were significantly higher for the combustion of rapeseed straw pellets, compared to the rates for woody pellets (Table 4). Available scientific publications report FE emission rates from the combustion of rapeseed straw to the order of 3700–13,000 mg·kg^{-1} [38], and from the combustion of wood pellets 430–1200 mg·kg^{-1} [39–41]. PM emission factors of biomass combustion expressed in units of mass per unit of energy are much more common in the literature. Chandrasekaran et al. [42] found that PM emissions resulting from grass pellet combustion were higher than received from woody pellets at both high (36–60 mg·MJ^{-1}) and low loads (26–40 mg·MJ^{-1}). In addition, the ash content of the fuel was strongly correlated with emissions of PM2.5. According to the literature reports [43], during the combustion of the currently used woody fuels, the amount of emitted particles ranged from 13 to 92 mg·MJ^{-1}. However, the combustion of grain biofuels (oat, rape seed, rape seed residue), in contrast to wood biofuels, resulted in significant emissions of the phosphate fraction of particulate matter. In continuous biomass-fired equipment, the content of alkali metals coming from the fuel is a major factor in particulate formation. Ozgen et al. [44], on the other hand, reported that in a study of boilers fired with various wood biofuels, PM emissions were significantly lower for automatic pellet-fired equipment than for manual (wood-fired) equipment at 85 g·GJ^{-1}, and the average particulate emission rates for wood pellets were 6–116 g·GJ^{-1} [45].

During the boiler tests conducted, high particle emissions associated with the combustion of rapeseed straw biomass pellets were found, which may discredit this biomass as a fuel for low-emission boilers. Consideration should therefore be given to the use of agrobiomass solid biofuels in larger capacity installations equipped with flue gas cleaning systems. Bearing in mind that pellet fuel generates lower particulate emissions than other types of wood fuels. According to [46], particulate emissions can be far from the emission limit when burning pellets made of wood in grate burners with electrostatic precipitators. Shen et al. [47] reported that biomass pellets can be a clean replacement for biomass in its traditional form. This indicates the need to study emissions from the combustion of biomass pellets.

Small-capacity boilers used for domestic heating are a significant source of particulate air pollution in the winter season. The efficiency of the dedusting devices should be higher than 95% to meet the ECOSOC emission limit and the emission level for a biomass lower than 20 mg·m^{-3}. Therefore, an urgent need has arisen to develop small flue gas cleaning devices dedicated for particulate matter produced by small domestic boilers, which could be integrated with such boilers [48].

3.2. Engine Test Results

During the testing of a John Deere engine fueled by FAME and comparatively by diesel fuel, the content of PM emissions in the exhausts was measured under varying load-speed conditions.

Figures 6–10 show the changes in the PM concentration in the engines' exhausts when operating over a range of load characteristics realized at 1400, 1650, 1900, 2150, and 2400 rpm, respectively.

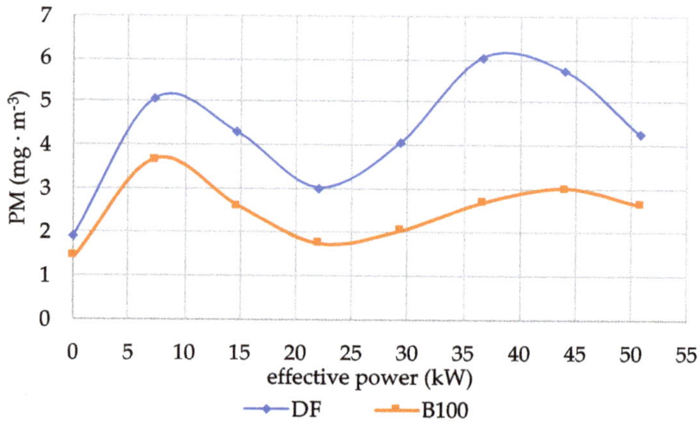

Figure 6. PM emissions vs. effective power of the John Deere 4045TF285JD engine at 1400 rpm.

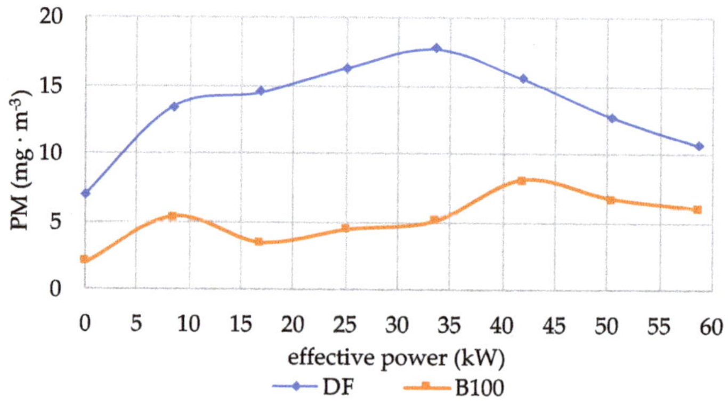

Figure 7. PM emissions vs. the effective power of the John Deere 4045TF285JD engine at 1650 rpm.

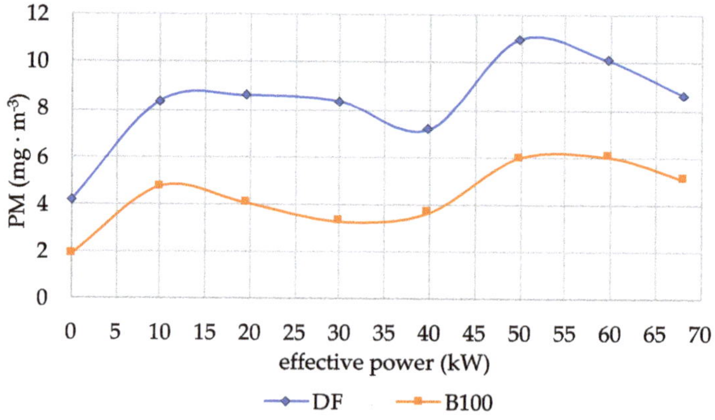

Figure 8. PM emissions vs. the effective power of the John Deere 4045TF285JD engine at 1900 rpm.

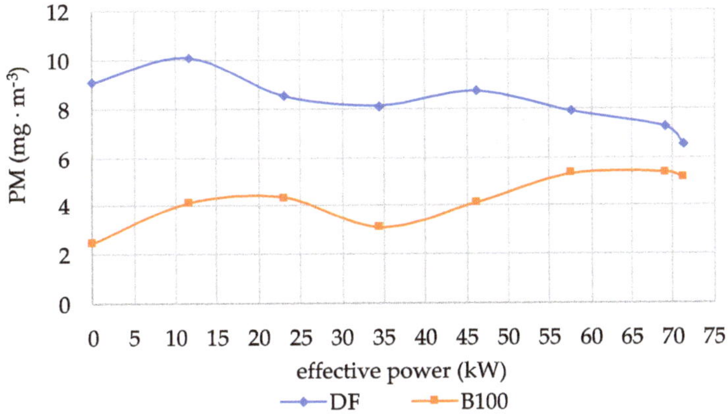

Figure 9. PM emissions vs. the effective power of the John Deere 4045TF285JD engine at 2150 rpm.

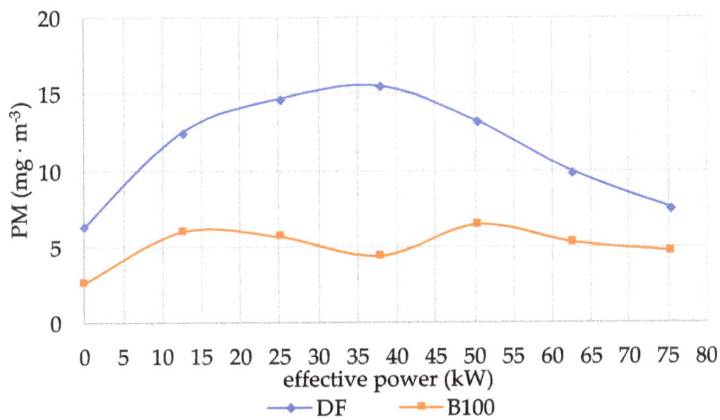

Figure 10. PM emissions vs. the effective power of the John Deere 4045TF285JD engine at 2400 rpm.

The tests showed significant reductions in PM emissions at all of the realized engine speeds for the biodiesel-fueled engine compared to the diesel-fueled one, averaging over 40% over the entire engine load range. The largest decrease in average PM emissions for biodiesel compared to diesel (61.1%) was recorded at 1650 rpm (i.e., corresponding to the highest engine torque (Figure 7)), and the smallest at 1400 rpm—40.3% (Figure 6). At a rated speed of 2400 rpm (Figure 10), the decrease in PM emissions of an engine fueled with B100 biofuel compared to a conventional fuel drive was 54.2% on average over the entire power range. At intermediate speeds of 1900 rpm (Figure 8) and 2150 rpm (Figure 8), the engine emitted less particulate matter when powered by biodiesel, averaging 48.2% and 46.9%, respectively, compared to DF. In the case of the B100 biofuel, for all rotational speeds, the lowest PM emission levels were observed for the operation of the no load engine operation. At the rated load for the considered speeds, the PM content in the exhaust gases of the engine fueled with biodiesel increased nearly three times. However, regardless of the load, the PM emissions for B100 each time were lower than that for the DF.

The significant decrease in the particulate matter content in the exhaust gases of the tested engine when fed with FAME was due to the high oxygen content in the biofuel (10.3%—Table 2). Such a significant oxygen share in the biofuel resulted in more complete combustion, which was bound with the reduced formation of PM.

In diesel engines, it is problematic to thoroughly mix the fuel with the oxygen contained in the air, which results in a local lack of oxygen and high-temperature breakdown of fuel particles, leading to the formation of particulate matter. Therefore, if part of the oxygen is supplied in the fuel, it will allow for more complete combustion and lower emissions of harmful exhaust components.

The significant environmental benefits of reducing diesel engine smoke when using rapeseed oil methyl esters have been confirmed in domestic and international studies. For example, the authors in [49,50] pointed to a nearly 50% reduction in PM emissions for the B100 biofuel compared to DF, confirming the downward trend in the emissions of this component in the studies conducted (a decrease of more than 40%). Other authors [51,52] have reported lower PM emissions for the B100 biofuel compared to DF in the range of 20–60%, which is also confirmed by the results of the studies included in this publication. However, the amount of particulate matter emitted is closely related to the specific engine (design features, technical condition, regulatory settings used, etc.) and the conditions under which it operates [53,54].

Table 5 presents the ANOVA results obtained for the measured PM emission levels in the boiler and engine tests.

Table 5. ANOVA results for PM emission levels ($mg \cdot m^{-3}$) due to fuel.

Fumes' Component	Factor	Degrees of Freedom df	Totals of Squares SS	Medium Square MS	Test Function Value F	Calculated Significance Level p
PM	Fuel wood pellets and rapeseed straw pellets)	1	18,659,683	18,659,683	161,346.3	0
PM	Fuel DF and B100	1	460.6794	460.6794	50.28063	0

Statistical studies were carried out to confirm the observed differences in the particulate emission levels. The results of the analysis of variance obtained for the PM emission levels by type of pellet showed significant differences between the average values (at the significance level of $\alpha = 0.05$).

Similarly for the pellets, the results of the analysis of variance obtained for the PM emission levels by the type of liquid fuel (B100, DF) showed significant differences between the average values (at the significance level of $\alpha = 0.05$).

The statistical evaluation of the significance of the differences (ANOVA) presented in Table 5 completes and makes the comparison of particulate matter emissions more credible due to the biofuel used in both energy devices.

4. Conclusions

The negative aspects of emissions associated with the combustion of various biofuels may prevent their use as sustainable fuels. To overcome this disadvantage, detailed information is needed on particulate emissions from the combustion of different types of biofuels. This information will help identify the types of biomass and biofuels that emit more particles during combustion and can lead to measures to reduce this pollution [29]. The properties and applications of different types of biofuels vary widely, as do their advantages and disadvantages [55]. These properties can significantly affect the air quality associated with combustion processes [56]. The purpose of the boiler tests conducted was to analyze particulate emissions during the combustion of wood pellets and waste biomass pellets—rapeseed straw—in a low-power boiler. The results obtained indicate that the level of particulate emissions accompanying the combustion of wood pellets was significantly lower compared to the level of emissions recorded during the combustion of the rapeseed pellets, and at this stage, wood pellets are by far the better biofuel for individual household use. However, rapeseed straw should not be discredited as a biofuel as it can be burned in higher-capacity installations equipped with flue gas cleaning devices. Further research into the combustion process of this biofuel may contribute to improving the emission rates. The development of waste biomass such as rapeseed straw will allow for the sustainable and efficient use of rapeseed crops and a closed material cycle in the economy. In addition, it is important to develop another line of research concerning the equipping of small domestic boilers with integrated devices designed to purify flue gases from particulate matter.

Engine tests conducted under varying speed and load conditions showed clear environmental benefits associated with significantly lower levels of particulate matter emissions in the engine exhaust fueled by biodiesel (40–60%) compared to diesel fuel. Such a large reduction in PM emissions makes rapeseed oil esters a desirable choice as an engine fuel in the agricultural sector, among others. This is due to the fact that tractor engines burn significant amounts of fuel. In addition, tractor engines often operate at high loads, often rated, under which operating conditions the concentration of particulate matter in the exhaust is high. In addition, the authors' studies have shown other measurable environmental benefits of biodiesel as a diesel fuel such as with regard to methane, whose emissions when running on biofuel compared to diesel fuel have been reduced by 25–30% [6].

Given the wide variety of biofuels, there is a need to continue research on their optimal use in specific equipment or energy processes with the least possible environmental impact.

Author Contributions: Conceptualization, J.W., J.S.-B. and G.Z.; Methodology, J.W. and G.Z.; Software, G.Z.; Validation, A.K. (Andrzej Kuranc); Formal analysis, J.W.; Investigation, J.W., G.Z. and J.S.-B.; Resources, J.W. and G.Z.; Data curation, G.Z.; Writing—original draft preparation, J.S.-B., J.W. and G.Z.; Writing—review and editing, A.K. (Andrzej Kuranc), A.K. (Adam Koniuszy), and M.H.-P.; Visualization, A.K. (Andrzej Kuranc); Supervision, A.K. (Andrzej Kuranc); Project administration, G.Z.; Funding acquisition, A.K. (Adam Koniuszy) and M.H.-P. All authors have read and agreed to the published version of the manuscript.

Funding: This research received no external funding.

Data Availability Statement: Not applicable.

Conflicts of Interest: The authors declare no conflict of interest.

References

1. Werle, S. *Termiczne Przetwarzanie Biomasy Odpadowej Jako Element Gospodarki Obiegu Zamkniętego*; Wydawnictwo Politechniki Śląskiej: Gliwice, Poland, 2021; ISBN 83-7880-764-9.
2. Rogowska, D. Wykorzystanie OZE w Energetyce a Zrównoważony Rozwój. *Nafta-Gaz* **2017**, *73*, 616–623. [CrossRef]
3. Dołżyńska, M.; Obidziński, S.; Piekut, J.; Yildiz, G. The Utilization of Plum Stones for Pellet Production and Investigation of Post-Combustion Flue Gas Emissions. *Energies* **2020**, *13*, 5107. [CrossRef]
4. Król, D.; Łach, J.; Poskrobko, S. O Niektórych Problemach Związanych z Wykorzystaniem Biomasy Nieleśnej w Energetyce. *Energetyka* **2010**, *1*, 53–62.
5. Juszczak, M. Concentrations of Carbon Monoxide and Nitrogen Oxides from a 15 KW Heating Boiler Supplied Periodically with a Mixture of Sunflower Husk and Wood Pellets. *Environ. Prot. Eng.* **2014**, *40*, 65–74. [CrossRef]
6. Wasilewski, J.; Zając, G.; Szyszlak-Bargłowicz, J.; Kuranc, A. Evaluation of Greenhouse Gas Emission Levels during the Combustion of Selected Types of Agricultural Biomass. *Energies* **2022**, *15*, 7335. [CrossRef]
7. Kachel, M.; Kraszkiewicz, A.; Subr, A.; Parafiniuk, S.; Przywara, A.; Koszel, M.; Zając, G. Impact of the Type of Fertilization and the Addition of Glycerol on the Quality of Spring Rape Straw Pellets. *Energies* **2020**, *13*, 819. [CrossRef]
8. Kraszkiewicz, A.; Santoro, F.; Pascuzzi, S. Emmission of Sulphur Oxides from Agricultural Solid Biofuels Combustion. *Agric. Eng.* **2020**, *24*, 35–45. [CrossRef]
9. Lech-Brzyk, K. Emisja wybranych zanieczyszczeń podczas spalania biomasy w kotłach o mocy 2÷4 MW. *Ochr. Środowiska* **2014**, *36*, 47–52.
10. Vicente, E.; Alves, C. An Overview of Particulate Emissions from Residential Biomass Combustion. *Atmos. Res.* **2018**, *199*, 159–185. [CrossRef]
11. Johansson, L.; Tullin, C.; Leckner, B.; Sjövall, P. Particle Emissions from Biomass Combustion in Small Combustors. *Biomass Bioenergy* **2003**, *25*, 435–446. [CrossRef]
12. Demirbas, A. Hazardous Emissions from Combustion of Biomass. *Energy Sources Part A Recovery Util. Environ. Eff.* **2007**, *30*, 170–178. [CrossRef]
13. Hawrot-Paw, M.; Koniuszy, A.; Sędłak, P.; Seń, D. Functional Properties and Microbiological Stability of Fatty Acid Methyl Esters (FAME) under Different Storage Conditions. *Energies* **2020**, *13*, 5632. [CrossRef]

14. Matwijczuk, A.; Zajac, G.; Kowalski, R.; Kachel-Jakubowska, M.; Gagos, M. Spectroscopic Studies of the Quality of Fatty Acid Methyl Esters Derived from Waste Cooking Oil. *Pol. J. Environ. Stud.* **2017**, *26*, 2643–2650. [CrossRef] [PubMed]
15. Badura, X. Profil Składu Chemicznego Cząstek Stałych (PM) Emitowanych Przy Zastosowaniu Paliw z Biokomponentami. *Nafta-Gaz* **2014**, *70*, 817–824.
16. Chłopek, Z. *Ekologiczne Aspekty Motoryzacji i Bezpieczeństwo Ruchu Drogowego*; Wydział Samochodów i Maszyn Roboczych Politechniki Warszawskiej: Warszawa, Poland, 2012.
17. Bowe, B.; Xie, Y.; Yan, Y.; Al-Aly, Z. Burden of Cause-Specific Mortality Associated with $PM_{2.5}$ Air Pollution in the United States. *JAMA Netw. Open* **2019**, *2*, e1915834.
18. Sharma, S.; Chandra, M.; Kota, S.H. Health Effects Associated with PM2. 5: A Systematic Review. *Curr. Pollut. Rep.* **2020**, *6*, 345–367. [CrossRef]
19. Yin, P.; Brauer, M.; Cohen, A.J.; Wang, H.; Li, J.; Burnett, R.T.; Stanaway, J.D.; Causey, K.; Larson, S.; Godwin, W. The Effect of Air Pollution on Deaths, Disease Burden, and Life Expectancy across China and Its Provinces, 1990–2017: An Analysis for the Global Burden of Disease Study 2017. *Lancet Planet. Health* **2020**, *4*, e386–e398. [CrossRef]
20. Alves, C.; Evtyugina, M.; Vicente, E.; Vicente, A.; Rienda, I.C.; de la Campa, A.S.; Tomé, M.; Duarte, I. $PM_{2.5}$ Chemical Composition and Health Risks by Inhalation near a Chemical Complex. *J. Environ. Sci.* **2023**, *124*, 860–874. [CrossRef]
21. EEA, E. *Air Quality in Europe—2019 Report*; European Environment Agency: Copenhagen, Denmark, 2019.
22. Šarkan, B.; Jaśkiewicz, M.; Kubiak, P.; Tarnapowicz, D.; Loman, M. Exhaust Emissions Measurement of a Vehicle with Retrofitted LPG System. *Energies* **2022**, *15*, 1184. [CrossRef]
23. Chłopek, Z.; Szczepański, T. Ocena Zagrożenia Środowiska Cząstkami Stałymi Ze Źródeł Cywilizacyjnych. *Inżynieria Ekol.* **2012**, *30*, 174–193.
24. MAHA MPM4 Partikelmessgerat. *Original-Betriebsanleitung*; MAHA Maschinenbau Haldenwang GmbH & Co. KG: Haldenwang, Germany, 2008.
25. Chojnacki, J.; Ondruška, J.; Kuczyński, W.; Šooš, L.; Bałasz, B. Emissions from the Combustion of Solid Biofuels. In Proceedings of the 9th International Scientific Symposium on Farm Machinery and Process Management in Sustainable Agriculture, Lublin, Poland, 22–24 November 2017; pp. 70–75.
26. Yang, W.; Zhu, Y.; Cheng, W.; Sang, H.; Yang, H.; Chen, H. Characteristics of Particulate Matter Emitted from Agricultural Biomass Combustion. *Energy Fuels* **2017**, *31*, 7493–7501. [CrossRef]
27. Carroll, J.; Finnan, J. Emissions and Efficiencies from the Combustion of Agricultural Feedstock Pellets Using a Small Scale Tilting Grate Boiler. *Biosyst. Eng.* **2013**, *115*, 50–55. [CrossRef]
28. Garcia-Maraver, A.; Zamorano, M.; Fernandes, U.; Rabaçal, M.; Costa, M. Relationship between Fuel Quality and Gaseous and Particulate Matter Emissions in a Domestic Pellet-Fired Boiler. *Fuel* **2014**, *119*, 141–152. [CrossRef]
29. Simões Amaral, S.; Andrade de Carvalho, J., Jr.; Martins Costa, M.A.; Pinheiro, C. Particulate Matter Emission Factors for Biomass Combustion. *Atmosphere* **2016**, *7*, 141. [CrossRef]
30. Limousy, L.; Jeguirim, M.; Dutournié, P.; Kraiem, N.; Lajili, M.; Said, R. Gaseous Products and Particulate Matter Emissions of Biomass Residential Boiler Fired with Spent Coffee Grounds Pellets. *Fuel* **2013**, *107*, 323–329. [CrossRef]
31. Rybak, W. *Spalanie i Współspalanie Biopaliw Stałych*; Oficyna Wydawnicza Politechniki Wrocławskiej: Wrocław, Poland, 2006; ISBN 83-7085-938-0.
32. Musialik-Piotrowska, A.; Kordylewski, W.; Ciołek, J.; Mościcki, K. Characteristics of Air Pollutants Emitted from Biomass Combustion in Small Retort Boiler. *Environ. Prot. Eng.* **2010**, *36*, 123–131.
33. Niu, Y.; Tan, H. Ash-Related Issues during Biomass Combustion: Alkali-Induced Slagging, Silicate Melt-Induced Slagging (Ash Fusion), Agglomeration, Corrosion, Ash Utilization, and Related Countermeasures. *Prog. Energy Combust. Sci.* **2016**, *52*, 1–61. [CrossRef]
34. Fournel, S.; Palacios, J.; Morissette, R.; Villeneuve, J.; Godbout, S.; Heitz, M.; Savoie, P. Influence of Biomass Properties on Technical and Environmental Performance of a Multi-Fuel Boiler during on-Farm Combustion of Energy Crops. *Appl. Energy* **2015**, *141*, 247–259. [CrossRef]
35. Vassilev, S.V.; Baxter, D.; Andersen, L.K.; Vassileva, C.G. An Overview of the Composition and Application of Biomass Ash: Part 1. Phase–Mineral and Chemical Composition and Classification. *Fuel* **2013**, *105*, 40–76. [CrossRef]
36. Vassilev, S.V.; Baxter, D.; Andersen, L.K.; Vassileva, C.G. An Overview of the Composition and Application of Biomass Ash: Part 2. Potential Utilisation, Technological and Ecological Advantages and Challenges. *Fuel* **2013**, *105*, 19–39. [CrossRef]
37. García, G.B.; de Hoces, M.C.; García, C.M.; Palomino, M.T.C.; Gálvez, A.R.; Martín-Lara, M.Á. Characterization and Modeling of Pyrolysis of the Two-Phase Olive Mill Solid Waste. *Fuel Process. Technol.* **2014**, *126*, 104–111. [CrossRef]
38. Wei, S.; Shen, G.; Zhang, Y.; Xue, M.; Xie, H.; Lin, P.; Chen, Y.; Wang, X.; Tao, S. Field Measurement on the Emissions of PM, OC, EC and PAHs from Indoor Crop Straw Burning in Rural China. *Environ. Pollut.* **2014**, *184*, 18–24. [CrossRef] [PubMed]
39. Schmidt, G.; Trouvé, G.; Leyssens, G.; Schönnenbeck, C.; Genevray, P.; Cazier, F.; Dewaele, D.; Vandenbilcke, C.; Faivre, E.; Denance, Y.; et al. Wood Washing: Influence on Gaseous and Particulate Emissions during Wood Combustion in a Domestic Pellet Stove. *Fuel Process. Technol.* **2018**, *174*, 104–117. [CrossRef]
40. Venturini, E.; Vassura, I.; Agostini, F.; Pizzi, A.; Toscano, G.; Passarini, F. Effect of Fuel Quality Classes on the Emissions of a Residential Wood Pellet Stove. *Fuel* **2018**, *211*, 269–277. [CrossRef]

41. Vicente, E.D.; Vicente, A.M.; Evtyugina, M.; Tarelho, L.A.C.; Almeida, S.M.; Alves, C. Emissions from Residential Combustion of Certified and Uncertified Pellets. *Renew. Energy* **2020**, *161*, 1059–1071. [CrossRef]
42. Chandrasekaran, S.R.; Hopke, P.K.; Hurlbut, A.; Newtown, M. Characterization of Emissions from Grass Pellet Combustion. *Energy Fuels* **2013**, *27*, 5298–5306. [CrossRef]
43. Sippula, O. Fine Particle Formation and Emissions in Biomass Combustion. *Rep. Ser. Aerosol Sci.* **2010**, *108*, 1–64.
44. Ozgen, S.; Caserini, S.; Galante, S.; Giugliano, M.; Angelino, E.; Marongiu, A.; Hugony, F.; Migliavacca, G.; Morreale, C. Emission Factors from Small Scale Appliances Burning Wood and Pellets. *Atmos. Environ.* **2014**, *94*, 144–153. [CrossRef]
45. Ozgen, S. Methods for Particulate Matter Emission Reduction from Pellet Boilers. *Biomass Convers. Biorefin.* **2022**. [CrossRef]
46. Ghafghazi, S.; Sowlati, T.; Sokhansanj, S.; Bi, X.; Melin, S. Particulate Matter Emissions from Combustion of Wood in District Heating Applications. *Renew. Sustain. Energy Rev.* **2011**, *15*, 3019–3028. [CrossRef]
47. Shen, H.; Luo, Z.; Xiong, R.; Liu, X.; Zhang, L.; Li, Y.; Du, W.; Chen, Y.; Cheng, H.; Shen, G.; et al. A Critical Review of Pollutant Emission Factors from Fuel Combustion in Home Stoves. *Environ. Int.* **2021**, *157*, 106841. [CrossRef] [PubMed]
48. Jaworek, A.; Sobczyk, A.; Marchewicz, A.; Krupa, A.; Czech, T. Particulate Matter Emission Control from Small Residential Boilers after Biomass Combustion. *A Review. Renew. Sustain. Energy Rev.* **2021**, *137*, 110446. [CrossRef]
49. Demirbas, A. Biofuels Securing the Planet's Future Energy Needs. *Energy Convers. Manag.* **2009**, *50*, 2239–2249. [CrossRef]
50. Maia, E.C.R.; Borsato, D.; Moreira, I.; Spacino, K.R.; Rodrigues, P.R.P.; Gallina, A.L. Study of the Biodiesel B100 Oxidative Stability in Mixture with Antioxidants. *Fuel Process. Technol.* **2011**, *92*, 1750–1755. [CrossRef]
51. Szlachta, Z. *Zasilanie Silników Wysokoprężnych Paliwami Rzepakowymi*; WkiŁ: Warszawa, Poland, 2002.
52. Swat, M.; Madej, K. The Effect of Bio-Components of Fuel on the Emission Character of Particulate Matter from Diesel Engines. *Probl. Eksploat.* **2012**, *3*, 203–212.
53. Arshad, M.; Zia, M.A.; Shah, F.A.; Ahmad, M. An Overview of Biofuel. In *Perspectives on Water Usage for Biofuels Production: Aquatic Contamination and Climate Change*; Arshad, M., Ed.; Springer International Publishing: Cham, Switzerland, 2018; pp. 1–37. ISBN 978-3-319-66408-8.
54. Silitonga, A.S.; Hassan, M.H.; Ong, H.C.; Kusumo, F. Analysis of the Performance, Emission and Combustion Characteristics of a Turbocharged Diesel Engine Fuelled with Jatropha Curcas Biodiesel-Diesel Blends Using Kernel-Based Extreme Learning Machine. *Environ. Sci. Pollut. Res.* **2017**, *24*, 25383–25405. [CrossRef]
55. Rafael, S.; Tarelho, L.; Monteiro, A.; Sá, E.; Miranda, A.; Borrego, C.; Lopes, M. Impact of Forest Biomass Residues to the Energy Supply Chain on Regional Air Quality. *Sci. Total Environ.* **2015**, *505*, 640–648. [CrossRef]
56. Garbaras, A.; Masalaite, A.; Garbariene, I.; Ceburnis, D.; Krugly, E.; Remeikis, V.; Puida, E.; Kvietkus, K.; Martuzevicius, D. Stable Carbon Fractionation in Size-Segregated Aerosol Particles Produced by Controlled Biomass Burning. *J. Aerosol Sci.* **2015**, *79*, 86–96. [CrossRef]

energies

Article

Urban Sustainability: Recovering and Utilizing Urban Excess Heat

Kristina Lygnerud [1,2,*] and Sarka Langer [1]

1 IVL Swedish Environmental Research Institute, SE-400 14 Göteborg, Sweden
2 Department of Energy Sciences, Efficient Energy Systems, Faculty of Engineering LTH, Lund University, SE-221 00 Lund, Sweden
* Correspondence: kristina.lygnerud@ivl.se or kristina.lygnerud@energy.lth.se; Tel.: +46-727-08-66-26

Abstract: Urban heat sources from urban infrastructure and buildings could meet ~10% of the European building heating demand. There is, however, limited information on how to use them. The EU project ReUseHeat has generated much of the existing knowledge on urban waste heat recovery implementation. Heat recovery from a data center, hospital and from water were demonstrated. Additionally, the project generated knowledge of stakeholders, risk profile, bankability and business models. The recovery of urban waste heat is characterized by high potential, high competitiveness compared to other heating alternatives, high avoidance of GHG emissions, payback within three years and low utilization. These characteristics reveal that barriers for increased utilization exist. The barriers are not technical. Instead, the absence of a waste heat EU level policy adds risk. Other showstoppers are low knowledge on the urban waste heat opportunity and new stakeholder relationships being needed for successful recovery. By combining key results and lessons learned from the project this article outlines the frontier of urban waste heat recovery research and practice in 2022.

Keywords: district heating; urban waste heat; demonstration sites; business aspects

check for updates

Citation: Lygnerud, K.; Langer, S. Urban Sustainability: Recovering and Utilizing Urban Excess Heat. *Energies* **2022**, *15*, 9466. https://doi.org/10.3390/en15249466

Academic Editors: Pandian Vasant, Vladimir Panchenko, Vadim Bolshev, Igor S. Litvinchev, Prasun Chakrabarti and Nallapaneni Manoj Kumar

Received: 26 October 2022
Accepted: 12 December 2022
Published: 14 December 2022

1. Introduction

The urban infrastructure of district heating (DH) is not new. The idea of DH is traced to ancient Roman baths. Early baths were heated using water from hot wells; later on, underfloor (hypocaust) heating was used e.g., a central heating system with an underground furnace where the hot air was distributed under a raised floor standing on pillars [1]. A precursor was established in the French village of Chaudes-Aigues in the 14th century. It consisted of wooden pipes distributing geothermal hot water from the hot spring of Par, with a temperature of 80–82 °C, to some buildings in the village [2,3]. The history of modern DH started in the United States in the middle of 18th century as single trials of private persons to heat their homes using combustion of wood, coal, oil or natural gas to produce high-temperature steam that was distributed through pipes [4]. In commercial form, DH has existed since the 1880s [5] and has constituted an urban infrastructure since. For example, the system supplying Manhattan in New York was put into operation in 1882. The current systems tend to have a supply temperature of approximately 80–90 °C, often referred to as third-generation systems [6]. In this paper, this kind of system is referred to as a high-temperature (HT) system. Low-temperature (LT) systems are increasingly relevant, as they allow for increased shares of renewables, geothermal and waste heat sources. They have been defined as systems with a supply temperature of heat that is below 70 °C [7]. Urban waste heat is LT, possible to introduce into both HT District Heating Networks (DHNs) and LTDHNs. When inserted into HTDHNs, a heat pump (HP) is often resorted to.

Half of the energy use in the EU is used for heating and cooling [8]. The total heat demand for buildings in Europe has been estimated at 10 EJ/yr [9]. Industrial waste heat (resulting from different processes, often HT) has a large potential to contribute to the energy demand in Europe. An estimated 2.7 EJ/yr of available industrial waste heat [10] could meet one quarter of the heat and hot water demand in Europe. Industrial waste heat has been successfully integrated into DHNs in some countries but there is still a large untapped potential. The world champion on industrial waste heat recovery into DHNs is Sweden, but even there only a fraction of heat supplied (9%) originates from industrial processes [11–17]. Urban waste heat can come from IT (data centers), transport systems (metro), sewage water and buildings. The ReUseHeat project identified that urban waste heat could satisfy one tenth of the European building heating demand.

Despite its potential, only a restricted number of installations are present in Europe. There are individual examples of heat recovery refrigeration system of supermarkets [18,19], from wastewater treatment facilities [20,21], from data centers [22–26] and from an underground station [27]. One explanation for the low implementation is that where HTDHNs exist, the interest in LT sources has been low as a result of other fuels such as gas and biomass being cost-efficient.

How warm a network is will determine whether urban heat sources need a heat pump to be recovered. In ReUseHeat, heat was recovered using an HP. In addition to technical validation, analyses were performed on potential, stakeholders, investment risk, bankability, contracts, business models and competitiveness compared to other heating alternatives. By combining key results and lessons learned from the project, this paper provides unique, holistic information on urban waste heat recovery. The results are aggregated and discussed jointly, providing information on the 2022 frontier of LT heat recovery research and practice.

In the context of EU-funded research, ReUseHeat (2017–2022) [28] builds on previous knowledge and EU-funded projects, focusing on things such as potential studies for DH including industrial HT waste heat recovery (the finalized Heat Roadmap Europe Series) and DH implementation to create awareness about the solutions at city level (the finalized CELSIUS Project) [29–31]. ReUseHeat bridged the gap between conventional HTDH and unconventional LTDH and has been followed by the ongoing REWARDHeat project [32] addressing standardized solutions for LT heat recovery.

Next, materials and methods applied for collecting different kinds of results in the ReUseHeat project are described. In Section 3, the results on urban waste heat potential, LTDH performance, barriers and business aspects are provided. Discussion (Section 4) and conclusions (Section 5) round the paper off.

2. Materials and Methods

The ReUseHeat project has demonstrated three demonstration sites recovering urban waste heat. Four sites were targeted, but one could not be implemented (metro system heat recovery). To contextualize the urban waste heat recovery, its potential was estimated for EU 28. Moreover, business aspects were studied in depth to support demonstrator replication as well as to create awareness about urban waste heat recovery characteristics. The work with demonstration, potential assessments and business aspects is heterogenous. Therefore, a multitude of methods were applied to generate different results. These are presented next.

2.1. EU Potential

The urban excess heat potential of waste heat encompassed sources other than those foreseen to be demonstrated in the project. Waste heat from datacenter, metro, hospital and water (which were to be demonstrated in ReUseHeat) and food processes and buildings were identified [33]. LT waste heat has the disadvantage that it cannot be transported very far. Therefore, only heat sources within 2 km of the existing DHN across the EU-28 countries, at an average HP performance of COP 3 of the HP used, were included in the final assessment of the accessible excess heat. How far the waste heat can be transported depends on the size of the source and how warm it is. It is therefore difficult to identify a cut-off distance that applies to all LT heat sources. A cutoff was made, allowing LT waste heat transportation for a maximum of 2 km. These volumes were referred to as accessible excess heat volumes, an important distinction to gross available heat volumes.

For quantification of excess heat from the urban sources, an inventory of unique district energy installations, data centers, metro stations, wastewater treatment plants, food production and retail facilities, service sector buildings and residential sector buildings was drawn up. The general basis for the accessible excess heat assessment for the service sector buildings and residential sector buildings was data on specific cooling demand (cooling need reflects available waste heat) [34]. The excess heat sources have been characterized by recovery type, temperate ranges, temporality and heat pump conversion type (Table S1 in Supplementary Material).

2.2. Method to Assess Scalability and Replicability

Scalability reflects how well a system, network or process can expand to meet increasing demand. Replicability indicates how well a system can be copied and installed somewhere else. The methodology for assessment of the scalability and replicability consisted of collecting specific data from demonstration sites by means of a questionnaire survey. Several factors were assessed to identify the scalability and replicability of the demonstrator sites. Economical, regulatory, and stakeholder acceptance are examples of factors assessed. A cumulative result—a scalability index and a replicability index—were calculated for each of the demonstration sites [35].

2.3. Method to Compare Costs of Alternatives for Heating

A calculation tool has been developed by ReUseHeat that compares the cost of LT heat recovery with alternative heating solutions. The levelized cost of LTDH was identified. For assumptions of the tool and details on calculations, see Supplementary Material and Tables S2–S4. The tool is downloadable from the webpage of the project.

2.4. Method to Study Business Aspects

The project identified the key stakeholders, barriers, value chain, risks, bankability, organization, contractual factors and business models. The stakeholder perspectives, barriers and the status of the value chain were studied: the scientific literature and the existing laws, policies, regulations and guidelines (collectively defined as 'institutional barriers') in Europe were reviewed; interviews with multiple stakeholders were held, involving 76 respondents across eight European countries. The stakeholder groups interviewed were DH companies, waste heat owners, customers, policy makers and investors interested in green energy. For the risk assessment, scenario analysis in combination with a discourse on cognitive bias were applied to the context of the demonstrators. For the bankability assessment, financial principles were applied to urban waste heat recovery investment opportunities. The contract design was based on traditional methodologies related to infrastructure projects [36]. For identifying efficient business models for the demonstrator sites, the business model canvas was used [37].

2.5. Method for Technical Demonstration

The technical demonstration was conducted stepwise;a pre-feasibility study was succeeded by a feasibility study, commissioning and subsequent operation. The progression of the demonstration sites was followed up on a quarterly basis. Once the equipment was taken into operation the results of the demonstrators were monitored. Performance data from the demonstration sites are: heat supply, excess heat saved, electricity, primary energy saved, CO_2 emissions saved and economic parameters such as simplified payback period. For the datacenter heat recovery, four months of monitored data were generated in the project. For the hospital heat recovery, 10 months of monitored data were generated in the project. Extrapolations for full-year operations were made for both demonstrators. For the awareness-generating demonstrator site, more than one year of monitoring data exists.

3. Results

The results are provided on urban waste heat potential (3.1), LT DH performance and barriers (3.2) and business aspects (3.3).

3.1. *Urban Waste Heat Potential*

The accessible waste heat volumes in the EU-28 countries are summarized in the supplementary materials (Table S5).

The total volume of the accessible urban waste heat is 1.2 EJ/yr per year. Most comes from sewage water (42%), followed by data centers (23%), buildings (service sector 19% and residential sector 8.8%). Smaller fractions come from metro systems and food processes. LT waste heat could meet ~10% of the European heat demand for buildings [9].

3.2. *Urban Waste Heat Recovery Performance*

3.2.1. Demonstrator Performance

Detailed information on demonstrators' concepts is described in Chapter 3: ReUseHeat handbook [38]. Several Key Performance Indictors (see Tables 7 and 9 in the ReUseHeat Handbook, Chapter 3 for specification of the KPIs) were quantified for the data center and a service sector building (hospital) demonstrator sites.

Data Center

The demonstrator is situated in Braunschweig (Germany). The heat is injected into a newly built and operated LTDHN. The performance is shown in Table 1.

Table 1. Intended and achieved key performance indicators data from the data center demonstrator.

Impact	Unit	Intended Result	Estimated Value for a Full Year
Heat supply	MWh/year	2300	2451
Excess heat volume	MWh/year	1750	1660
Electricity	MWh/year	580	791
Primary energy saved (PES)	MWh/year	1284	2602
CO_2 emissions saved	Tonnes/year	304	412
Simplified payback period	Years	8	3.1

Comparing the estimated results for a complete year with the intended values shows a large positive deviation in PES (doubled). CO_2 emissions saved and electricity usage were both larger than foreseen. More electricity was needed because hydraulic adjustments were necessary to avoid overheating of the HP. In terms of economic indication, the payback for results of a full year is foreseen to be much lower than anticipated (3.05 years instead of eight years).

Heat Recovery from a Hospital

The hospital is a public hospital in Madrid, Spain. LT heat from the condensation circuit of water-water electric chillers is recovered. The monitored data on the performance of the demonstrator are shown in Table 2.

The results show that the estimated results for a full year were better than expected. Again, the use of electricity was higher than foreseen but to be expected from the increased thermal energy production. Economically, the demonstrator had a significant shift of simplified payback from 15 to less than two years.

Table 2. Intended and achieved key performance indicators data from the data center demonstrator.

Impact	Unit	Intended Result	Estimated Value for a Full Year
Heat supply	MWh/year	770	2704
Waste heat recovered	MWh/year	532	1751
Electricity	MWh/year	238	789
Primary energy saved	MWh/year	554	3768
CO_2 emissions saved	Tonnes/year	154	721
Simplified payback period	Years	15	1.9

Awareness Building Demonstrator (Dashboard)

The third demonstration site was focused on developing a demonstrator for building awareness about the urban waste heat recovery by means of a visualization dashboard. To visualize is especially important in countries where heat demand provided by DH is low and awareness on DH is rather absent. It was developed for heat recovery from water (sea and sewage). The dashboard can be accessed through the project website. Detailed information is found in [39].

Metro Heat Recovery Demonstrator

A fourth demonstrator site was foreseen: metro tunnel/platform heat recovery. The metro demonstrator was not realized, as the stakeholders withdrew from the project. In spite of this, learnings were made from feasibility studies. Two design concepts exist and are ready for future implementation. For a review of these, please view Chapter 3 in the ReUseHeat Handbook [38].

Scalability and Replicability Analysis of Demonstrator Sites

The aggregated scalability and replicability indices of the demonstrators are presented in Figure 1. Both ratios were above 50% (the red dashed line) for all demonstrators. The scalability index was highest for the hospital demonstrator (75) and the replicability index was highest for the metro demonstrator (80). The scores for the individual factors of scalability and replicability of all four demonstrators are presented in Figures S1 and S2 in Supplementary Material.

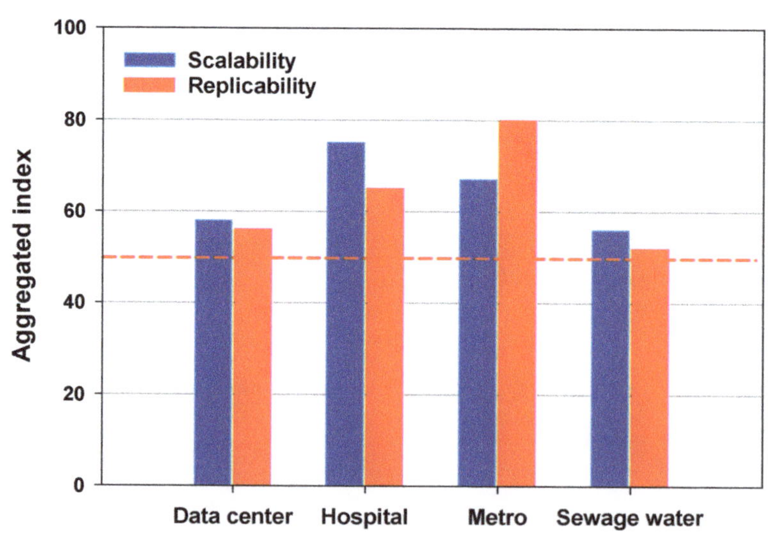

Figure 1. Scalability and replicability indexes by heat source.

The detailed analysis of the scalability and replicability factors identified economy of scale as the most important factor for all demonstrator sites. Profitability was also an important factor for three of them (except for the hospital demonstrator). Software integration, interface design and technical development were the low-score scalability factors.

3.2.2. Competitiveness of Urban Waste Heat Recovery

A tool was developed in the project to compare the costs of heating alternatives. It was applied to the following heat supply options: gas-, biomass-, oil- and electric boilers, air-to-water heat pump, HT DH and LT DH in the ReUseHeat demonstrator countries (Spain, France and Germany). In Germany, resorting to HT DH (€83/MWh) or LT DH (€74/MWh) are on par with the costs of a gas boiler (€83/MWh). The pattern is similar for a house in Spain, with the LCOHs being €77/MWh, €65/MWh and €80/MWh, respectively for the HT DH system, LT DH system and natural gas-fired boiler. The highest LCOHs in both countries are associated with electric boilers. In France, on the other hand, the natural gas-fired boiler is the cheapest alternative (€80/MWh) compared to the DH solutions (HT DH system: €98/MWh; LT DH system: €89/MWh). Detailed results with a distribution of the individual cost types are presented in Supplementary Material (Figures S3–S5) and at the website of the project (2021 energy prices used).

3.3. Barriers

Policy: In the EU, there is no clarity on what waste heat is. There is no policy setting waste heat on par with, for example, solar or wind. Instead, there are incentivized investments in renewables which then compete with non-incentivized investments in urban excess heat. The unclear status of excess heat adds investment risk to any waste heat recovery investment (urban and industrial).

System maturity: Incomplete value chains and limited demonstration makes it difficult to both find competencies that can install the system for heat recovery and to make standardized implementations. Instead, every time a new design is needed, and installers face a learning curve during the implementation. This is also reflected in the fact that there are no standards to adhere to and no standardized contracts to resort to. Jointly, these aspects make the implementation more costly and time-consuming than conventional DH system implementations.

Value of waste heat: Another barrier beyond the institutional involves different perceptions of how much the waste heat is worth. This is particularly troublesome if the parties have different expectations of, for example, payback of investments. Also, there might be different views on the quality and usefulness of the heat. For example, DH companies often do not need waste heat in summer regardless of how high its quality is, which leads to different perspectives of the value of the excess resource across seasons [40].

3.4. Results on Business Aspects

Stakeholders and value chain: The main stakeholders have been identified [41,42]: DH companies, excess-heat owners, customers, investors and policymakers.

The LT value chain is not complete and piggybacks off the high-temperature DH value chain. DH companies are interested in completing the parts missing to make LT heat recovery profitable. The waste heat owners are important for the success of LT heat recovery but must be willing to engage in contracts delivering specified heat volumes over time. At the side of the value chain are investors and policy makers. They can impact demand and support market uptake by demanding and incentivizing the LT heat recovery solution.

Contracts and risks: Regarding the contracts for urban waste heat investments, the project often requires multiple parties, which makes the contract writing complex. In designing contracts, important factors are win-win solutions, supply conditions, ownership and usage of assets, clear communication pathways, operational activities, renegotiation, mitigation and simplicity of the contracts [36,40,43]. The question of contractual efficiency was addressed in [36].

Business models: Working on business models for the ReUseHeat demonstrators, a transition from the business model logic of centralized and large-scale thinking to also include the value that a local heat source can offer was identified as important. The urban waste heat offer is characterized by the customer value of sustainably and locally produced. On the activity side, the urban waste heat recovery relies on long-term, win-win relationships with heat owners, often prosumers, which necessitates ample customer dialogue. On the resource side, the inclusion of an HP and possibly a LTDHN must be accounted for. One important learning from the project was that the sustainability that customers recognize is not capitalized on. Instead, the conventional HT business model is applied to the LT business case, which erodes it. Indeed, the sustainable feature of LT heat recovery could be an opportunity for DH companies to diversify their customer offer.

4. Discussion

The urban waste heat potential is of such magnitude that it should be a heat supply worth pursuing. Taking into account that it also has features that will be standard in future energy supply (no combustion, making use of a local resource in a circular system) and that it can replace fossil fuels, it should be on the agenda of any urban development scheme.

The urban heat sources will differ in terms of how large they are and how warm they are. The larger and warmer, the further the heat can be transported before use. The main delimitation is that LT heat must be used near where it is generated, as transportation or long supply lines are not efficient. This makes the matching between demand and supply increasingly important compared to a conventional HTDHN. If there is not enough demand locally for the available LT heat, then there are limited possibilities to use the full LT heat volumes available. This was, for example, the case of the datacenter heat recovery and for the foreseen implementation of the metro heat recovery.

The demonstrated site of datacenter heat recovery and heat recovery from cooling towers of a hospital show important results. Primary energy savings for a year from those two demonstrator sites is 6.3 GWh, and 1133 tonnes of CO_2 are saved; this is possible within a payback of three years (3.05 for the datacenter) or less (1.9 for the hospital). To put the size of the saving into proportion, an average-sized electric car uses 2 kWh per 10 km. The circumference at the equator is 40,074 km and to drive around it (theoretically) in the electric car one would need 2 kWh $\times$ 4007.4, which equals 8015 kWh or 0.008 GWh. Hence, the primary energy saved would allow an electrical car to drive 788 laps around the equator. For the context of the GHG savings, one ton of CO_2 emissions corresponds to using a hair dryer for 20,000 h. The tonnes of CO_2 saved would allow the usage of a hair dryer for 22.6 million hours or 944,167 days. The payback result was not expected. Rather, at the beginning of the project the novelty of the implementations and the absence of standards led to the assumption of paybacks in the range of 8–15 years. At the beginning of the project, the pre-assumption was also the LTDH solutions would have difficulty competing with gas boilers. For both Spain and Germany, applying the prices of 2021 (e.g., prices before the Russia-Ukraine war situation), LTDH proved to be a competitive option.

Profitability and a certain volume (scale) of the implementations were seen as important for scalability and replicability of the sites, whereas software integration for efficient operation was not seen as an issue. The most scalable site was the heat recovery from the hospital, whereas the most replicable site was the foreseen metro heat recovery. It was foreseen from the tunnel and platforms in metro systems. This implementation was the second foreseen implementation in Europe. In the CELCIUS Project, heat recovery was installed in the station of Islington in the metro system of London. Heat was recovered from the ventilation shafts and from transformers of electricity substations. The demonstrator encountered a number of barriers to implementation; one important one was the need to rebuild existing infrastructure to recover the waste heat. The ReUseHeat demonstrator foreseen for the metro system took this experience into account, and it was decided to target the heat from tunnels and station platforms. The idea was to make a compact implementation that could be placed in any metro tunnel. Returning to the element of distance, the distance between the heat recovery foreseen (in between rails in a small platform) and the customer (the building of the metro operator itself) complicated the implementation of the metro heat recovery. The project reviewed three alternative implementation sites for the metro heat recovery, where one would have been very efficient in terms of distance between heat supply and heat usage. This site had to be abandoned, as the metro operator decided to rebuild the space where the HP was foreseen to be installed.

That waste heat owners have core business activities that reduce the interest in waste heat recovery is known already from industrial waste heat collaborations. This was, however, also confirmed by the experiences in ReUseHeat, and it has been concluded that for new processes to be tested, organizational approvals take a long time. Several obstacles are identified and indicate that large-scale implementation will not come without an important effort. Some activities could support the development: (i) establishing that LT excess heat is a valuable asset at EU level and pushing its implementation by public sector requirements for urban waste heat recovery in new development areas; (ii) strengthening knowledge about the hidden urban asset; when there is awareness across the value chain from policy makers to customers, demand will follow; (iii) ensuring that waste heat investments are supported and placing them on a level playing field with investments in renewables; the current situation might lead to locally available heat supply being foregone; (iv) more implementation is needed to show the viability of urban heat recovery solutions. Standardization of technical configuration as well as of contractual arrangements are still pending. Not until such are in place will there be any large-scale private investment in this asset.

The DH market is heterogenous across countries. In addition, an EU-level framework on waste heat is missing, which makes it difficult for urban waste heat investments to keep pace with incentivized investments in renewable sources. Taking its large potential into account, it is important to foster interest in urban waste heat at both national and local levels. One possible way to push implementation is to make urban waste heat recovery standard in the construction of public spaces such as schools, hospitals and offices. Thereto, making heat planning mandatory at the municipal level across the EU would be feasible.

5. Conclusions

Globally, LT heat recovery has been implemented in a large number of places (more than 160 have been documented) [7], now augmented by the achievements from the ReUseHeat project. The number of these smart city installations confirms that interest in LT heat recovery is global. ReUseHeat project results validate that recovery of urban waste heat is technically, economically and environmentally feasible and can significantly support the decarbonization of cities [44,45].

In sum, the technology is there, and the heat supply is there; however, the policy framework and awareness amongst stakeholders are not. As a result, the demand is limited, and actors across the DH value chain deliver solutions they are used to delivering. In the light of the climate crisis and the Russia-Ukraine war, a strategy of "keeping the lights on" is no longer justifiable. It seems as if the time for large-scale LT heat recovery implementations has come.

Supplementary Materials: The following supporting information can be downloaded at: https://www.mdpi.com/article/10.3390/en15249466/s1, Figure S1: Computed Scalability factors; Figure S2: Computed Replicability factors; Figure S3: The LCOH estimations for the analysed heat supply options for Germany; Figure S4: The LCOH estimations for the analysed heat supply options for Spain; Figure S5: The LCOH estimations for the analysed heat supply options for France; Table S1: Excess heat source types, recovery types, temperature ranges, temporality and the HP conversion type for the investigated heat sources; Table S2: The techno-economic parameters assumed for the LCOH calculation of the individual and DH technologies—Germany; Table S3: The techno-economic parameters assumed for the LCOH calculation of the individual and DH technologies—Spain; Table S4: The techno-economic parameters assumed for the LCOH calculation of the individual and DH technologies—France; Table S5: Sources of urban excess heat, number of source units within the distance of two kilometers from a DHN and energy data (in the unit of PJ/year). References [46–62] are cited in the supplementary materials.

Author Contributions: K.L.: Ideation, Linking the results together from various deliverables in the project to provide a holistic perspective of LTDH, Funding acquisition, Project administration, Writing—Original Draft. S.L.: Review of material and analysis, Drafting, Review. All authors have read and agreed to the published version of the manuscript.

Funding: Materials collected from 'ReUseHeat', EH U2020, Grant Agreement 767429.

Data Availability Statement: Data can be obtained if requested.

Acknowledgments: We thank all contributors to ReUseHeat deliverables resorted to for creating this state of the art summary of urban waste heat recovery research and practice in 2022.

Conflicts of Interest: The authors declare no conflict of interest.

List of Abbreviations

DH	District Heating
DHN	District Heating Network
LT	Low Temperature
HT	High Temperature
HP	Heat Pump

References

1. Roman Baths. Available online: https://www.worldhistory.org/Roman_Baths/ (accessed on 26 July 2022).
2. Raynal, P.; Gibert, J.; Barthomeuf, C. Chaudes-Aigues: Historique des utilisations de la géothermie. *Reseaux Chal.* **1992**, *4*, 67–75.
3. Chaudes-Aigues: France's First Heating Network. Available online: https://www.dhcnews.net/chaudes-aigues-frances-first-heating-network/ (accessed on 26 July 2022).
4. Collins, J.F., Jr. The history of district heating. *Dist. Heat.* **1959**, *44*, 154–161.
5. Werner, S. *Development and Spread of District Heating (in Swedish: Fjärrvärmens Utveckling Och Utbredning)*; Värmeverksföreningen: Stockholm, Sweden, 1989.
6. Frederiksen, S.; Werner, S. *District Heating and Cooling*, 1st ed.; Studentlitteratur AB: Lund, Sweden, 2013; ISBN 9789144085302.
7. Lygnerud, K.; Werner, S. Low-Temperature District Heating Implementation Guidebook, IEA-DHC, Annex TS2, Final Report. 2021. Available online: https://www.iea-dhc.org/the-research/annexes/2017-2021-annex-ts2 (accessed on 26 July 2022).
8. Köhler, B.; Dengler, J.; Dinkel, A.; Mauman, A.; Kalz, D.; Bonato, P.; Fleitern, T.; Steinbach, J.; Ragwitz, M.; Arens, M.; et al. Mapping and Analyses of the Current and Future (2020–2030) Heating/Cooling Fuel Deployment (Fossil/Renewables). Work package 2: Assessment of the technologies for the year 2012. *Technology* **2016**. [CrossRef]
9. Werner, S. International review of district heating and cooling. *Energy* **2017**, *137*, 617–631. [CrossRef]
10. Miró, L.; Brückner, S.; Cabeza, L.F. Mapping and discussing Industrial Waste Heat (IWH) potentials for different countries. *Renew. Sustain. Energy Rev.* **2015**, *51*, 847–855. [CrossRef]
11. Bühler, F.; Petrović, S.; Karlsson, K.; Elmegaard, B. Industrial excess heat for district heating in Denmark. Appl. *Energy* **2017**, *205*, 991–1001. [CrossRef]
12. Fang, H.; Xia, J.; Zhu, K.; Su, Y.; Jiang, Y. Industrial waste heat utilization for low temperature district heating. *Energy Policy* **2013**, *62*, 236–246. [CrossRef]
13. Sun, F.; Cheng, L.; Fu, L.; Gao, J. New low temperature industrial waste heat district heating system based on natural gas fired boilers with absorption heat exchangers. *Appl. Therm. Eng.* **2017**, *125*, 1437–1445. [CrossRef]
14. Andrés, M.; Regidor, M.; Macía, A.; Vassalo, A.; Lygnerud, K. Assessment methodology for urban excess heat recovery solutions in energy-efficient District Heating Networks. *Energy Procedia* **2018**, *149*, 39–48. [CrossRef]
15. Nielsen, S.; Hansen, K.; Lund, R.; Moreno, D. Unconventional Excess Heat Sources for District Heating in a National Energy System Context. *Energies* **2021**, *13*, 5068. [CrossRef]
16. Soloha, R.; Pakere, I.; Blumberga, D. Solar energy use in district heating systems. A case study in Latvia. *Energy* **2017**, *137*, 586–594. [CrossRef]
17. Oktay, Z.; Aslan, A. Geothermal district heating in Turkey: The Gonen case study. *Geothermics* **2007**, *36*, 167–182. [CrossRef]
18. Sawalha, S. Investigation of heat recovery in CO_2 trans-critical solution for supermarket refrigeration. *Int. J. Refrig.* **2013**, *36*, 145–156. [CrossRef]
19. Zühlsdorf, B.; Christiansen, A.R.; Holm, F.M.; Funder-Kristensen, T.; Elmegaard, B. Analysis of possibilities to utilize excess heat of supermarkets as heat source for district heating. *Energy Procedia* **2018**, *149*, 276–285. [CrossRef]
20. Somogyi, V.; Sebestyén, V.; Domokos, E. Assessment of wastewater heat potential for district heating in Hungary. *Energy* **2018**, *163*, 712–721. [CrossRef]

21. Spriet, J.; McNabola, A.; Neugebauer, G.; Stoeglehner, G.; Ertl, T.; Kretschmer, F. Spatial and temporal considerations in the performance of wastewater heat recovery systems. *J. Clean. Prod.* **2020**, *247*, 119583. [CrossRef]
22. Huang, P.; Copertaro, B.; Zhang, X.; Shen, J.; Löfgren, I.; Rönnelid, M.; Fahlen, J.; Andersson, D.; Svanfeldt, M. A review of data centers as prosumers in district energy systems: Renewable energy integration and wasteheat reuse for district heating. *Appl. Energy* **2020**, *258*, 114109. [CrossRef]
23. Oró, E.; Taddeo, P.; Salom, J. Waste heat recovery from urban air cooled data centres to increase energy efficiency of district heating networks. *Sustain. Cities Soc.* **2019**, *45*, 522–542. [CrossRef]
24. Petrović, S.; Colangelo, A.; Balyk, O.; Delmastro, C.; Gargiulo, M.; Simonsen, M.B.; Karlsson, K. The role of data centres in the future Danish energy system. *Energy* **2020**, *194*, 116928. [CrossRef]
25. Wahlroos, M.; Syri, S.; Pärssinen, M.; Manner, J. Utilizing data center waste heat in district heating—Impact on energy efficiency and prospects for low-temperature district heating networks. *Energy* **2017**, *140*, 1228–1238. [CrossRef]
26. Wahlroos, M.; Pärssinen, M.; Rinne, S.; Syri, S.; Manner, J. Future views on waste heat utilization—Case of data centers in Northern Europe. *Renew. Sustain. Energy Rev.* **2018**, *82*, 1749–1764. [CrossRef]
27. Davies, G.; Boot-Handford, N.; Curry, D.; Dennis, W.; Ajileye, A.; Revesz, A.; Maidment, G. Combining cooling of underground railways with heat recovery and reuse. *Sustain. Cities Soc.* **2019**, *45*, 543–552. [CrossRef]
28. ReUseHeat—Recovery of Urban Excess Heat. European Commission. Grant Agreement Number: 767429. H2020-EE-2016-2017/H2020-EE-2017-RIA-IA. Available online: https://www.reuseheat.eu/ (accessed on 20 July 2022).
29. Nijs, W.; Castelló, P.R.; González, I.H. Heat Roadmap Europe. Baseline Scenario of the Total Energy System up to 2050. 2017. Available online: https://heatroadmap.eu/wp-content/uploads/2018/11/HRE4_D5.2.pdf (accessed on 26 July 2022).
30. HRE3/Stratego. 2022. Available online: https://heatroadmap.eu/sp_faq/heat-roadmap-europe-3-stratego-2015/ (accessed on 26 July 2022).
31. Celsius. Celsius-Smart Cities. 2020. Available online: https://celsiuscity.eu/ (accessed on 26 July 2022).
32. REWARDHeat. 2022. Available online: https://www.rewardheat.eu/en/ (accessed on 26 July 2022).
33. Persson, U.; Atabaki, S.; Nielsen, S.; Moreno, D. Report on the Amounts of Urban Waste Heat Accessible in the EU28: Update of Deliverable 1.4. Deliverable 1.9. 2022. Available online: https://www.reuseheat.eu/wp-content/uploads/2022/09/D1.9-Report-on-amounts-of-urban-waste-heat-accessible-in-EU28.pdf (accessed on 14 October 2022).
34. Persson, U.; Averfalk, H.; Nielsen, S.; Moreno, D. Accessible Urban Waste Heat. Deliverable 1.4 (Revised version). ReUseHeat. Recovery of Urban Excess Heat. 2020. Available online: https://www.reuseheat.eu/wp-content/uploads/2021/02/D1.4-Accessible-urban-waste-heat_revised-compressed.pdf (accessed on 20 July 2022).
35. Leonte, D. Scalability, Replicability and Modularity. Deliverable 2.9. ReUseHeat. Recovery of Urban Excess Heat. 2021. Available online: https://www.reuseheat.eu/wp-content/uploads/2021/09/D2.9-Scalability-replicability-and-modularity_Final-version_April-2021.pdf (accessed on 22 September 2022).
36. Wheatcroft, E.; Wynn, H.P.; Volodina, V.; Dent, C.J.; Lygnerud, K. Model-Based Contract Design for Low Energy Waste Heat Contracts: The Route to Pricing. *Energies* **2021**, *14*, 3614. [CrossRef]
37. Ostewalder, A.; Pigneur, Y. *Business Model Generation*; Wiley: New York, NY, USA, 2010; ISBN 978-0-470-87641-1.
38. Lygnerud, K.; Nielsen, S.; Persson, U.; Wynn, H.; Wheatcroft, E.; Antolin-Gutierrez, J.; Leonte, D.; Rosebrock, O.; Ochsner, K.; Keim, C.; et al. *Handbook for Increased Recovery of Urban Excess Heat*; Deliverable 6.2. ReUseHeat. Recovery of Urban Excess Heat; European Commission: Brussels, Belgium, 2022; ISBN 978-91-7883-404-4. Available online: https://www.euroheat.org/static/378761b4-2d76-48ef-a77a73730832b05a/ReUseHeat-Handbook-For-Increased-Recovery-of-Urban-Excess-Heat.pdf (accessed on 11 December 2022).
39. Antolín, J.; Sanz, R.; Miguel, F. Evaluation. Deliverable 4.5. ReUseHeat. Recovery of Urban Excess Heat. 2022. Available online: www.reuseheat.eu (accessed on 11 December 2022).
40. Lygnerud, K.; Wheatcroft, E.; Wynn, H. Contracts, Business Models and Barriers to Investing in Low Temperature District Heating Projects. *Appl. Sci.* **2019**, *9*, 3142. [CrossRef]
41. Leonte, D. Market and Stakeholder Analysis. Deliverable 2.1. ReUseHeat. Recovery of Urban Excess Heat. 2019. Available online: https://www.reuseheat.eu/wp-content/uploads/2019/03/D2.1-Market-and-stakeholder-analysis.pdf (accessed on 22 September 2022).
42. Wheatcroft, E.; Wynn, H.; Lygnerud, K.; Bonvicini, G.; Leonte, D. The Role of Low Temperature Waste Heat Recovery in Achieving 2050 Goals: A Policy Positioning Paper. *Energies* **2020**, *13*, 2107. [CrossRef]
43. Wynn, H.; Wheatcroft, E.; Lygnerud, K. Efficient Contractual Forms and Business Models for Urban Waste Heat Recovery. Deliverable 2.3. ReUseHeat. Recovery of Urban Excess Heat. 2021. Available online: https://www.reuseheat.eu/wp-content/uploads/2021/03/D2.3-UPDATED_20210223.pdf (accessed on 22 September 2022).
44. European Commission. Committing to Climate-Neutrality by 2050: Commission Proposes European Climate Law and Consults on the European Climate Pact. 2020. Available online: https://ec.europa.eu/commission/presscorner/detail/en/ip_20_335 (accessed on 26 July 2022).
45. European Union. A European Green Deal. 2021. Available online: https://ec.europa.eu/info/strategy/priorities-2019-2024/european-green-deal_en (accessed on 26 July 2022).
46. IEA—International Energy Agency. World Energy Model. Macro Drivers. Available online: https://www.iea.org/reports/world-energy-model/macro-drivers (accessed on 26 July 2022).

47. Danish Energy Agency. Technology Data. Available online: https://ens.dk/en/our-services/projections-and-models/technology-data (accessed on 26 July 2022).
48. Hansen, C.H.; Gudmundsson, O. The-Competitiveness-of-District-Heating-Compared-to-Individual-Heatingv2. 2018. Available online: https://www.danskfjernvarme.dk/-/media/danskfjernvarme/gronenergi/analyser/03052018-the-competitiveness-of-district-heating-compared-to-individual-heatingv2.pdf (accessed on 26 July 2022).
49. Greenhouse Gas Emission Intensity of Electricity Generation in Europe. Available online: https://www.eea.europa.eu/ims/greenhouse-gas-emission-intensity-of-1 (accessed on 26 July 2022).
50. Schüppler, S.; Fleuchaus, P.; Blum, P. Techno-economic and environmental analysis of an Aquifer Thermal Energy Storage (ATES) in Germany. *Geotherm. Energy* **2019**, *7*, 11. [CrossRef]
51. Malaich, B.; Oschatz, B. BDEW-Heizkostenvergleich Altbau 2021. Ein Vergleich der Gesamtkosten verschiedener Systeme zur Heizung und Warmwasserbereitung in Altbauten. 2021. Available online: https://www.bdew.de/media/documents/BDEW-HKV_Altbau.pdf (accessed on 26 July 2022).
52. Großklos, M. Kumulierter Energieaufwand und CO2-Emissionsfaktoren Verschiedener Energieträger und–Versorgungen. 2020. Available online: https://www.iwu.de/fileadmin/tools/kea/kea.pdf (accessed on 26 July 2022).
53. Miara, M.; Günther, D.; Kramer, T.; Oltersdorf, T.; Wapler, J. Wärmepumpen effizienz. Messtechnosche Untersuchung von Wärmepumpenangelan zur Analyse und Bewertung der Effizienz in realen Betrieb. 2011. Available online: https://wp-monitoring.ise.fraunhofer.de/wp-effizienz//download/wp_effizienz_endbericht_langfassung.pdf (accessed on 26 July 2022).
54. Zukunfsheizen. Available online: https://www.zukunftsheizen.de/brennstoff/zusammensetzung-heizoelpreis/ (accessed on 26 July 2022).
55. Bundesministerium für Wirtschaft und Klimaschuts. Staatlich veranlasste Bestanteile des Gaspreises. Available online: https://www.bmwi.de/Redaktion/DE/Artikel/Energie/gaspreise-bestandteile-staatlich.html#:~:text=Die%20Energiesteuer%20(Gassteuer)%20f%C3%BCr%20die,und%20flie%C3%9Ft%20in%20den%20Bundeshaushalt (accessed on 26 July 2022).
56. Eurostat—Statistics Explained. Electricity Prices for Household Consumers, First Half 2021 v5. Available online: https://ec.europa.eu/eurostat/statistics-explained/index.php?title=File:Electricity_prices_for_household_consumers,_first_half_2021_v5.png#file (accessed on 26 July 2022).
57. Eurostat—Statistics Explained. Electricity Prices for Household Consumers, First Half 2021 v1. Available online: https://ec.europa.eu/eurostat/statistics-explained/index.php?title=File:Natural_gas_prices_for_household_consumers,_first_half_2021_v1.png (accessed on 26 July 2022).
58. AGFW—Der Energieeffizienzverband für Wärme, Kälte und KWK. Available online: https://ag-energiebilanzen.de/mitglied/agfw-der-energieeffizienzverband-fuer-waerme-kaelte-und-kwk-e-v/ (accessed on 11 December 2022).
59. Resolution BOE-A-2020-11426. Ministerio para la Transición Ecológica y el Reto Demográfico. Resolución de 29 de septiembre de 2020, de la Dirección General de Política Energética y Minas, por la que se Publica la Tarifa de Último Recurso de gas Natural. Available online: https://www.boe.es/diario_boe/txt.php?id=BOE-A-2020-11426 (accessed on 26 July 2022).
60. Precios de los Derivados del Petróleo: España. Available online: https://datosmacro.expansion.com/energia/precios-gasolina-diesel-calefaccion/espana (accessed on 26 July 2022).
61. Precio neto de la Electricidad para uso Doméstico y uso Industrial—Euros/kWh. Available online: https://www.mincotur.gob.es/es-es/IndicadoresyEstadisticas/BoletinEstadistico/Energ%C3%ADa%20y%20emisiones/4_12.pdf (accessed on 26 July 2022).
62. ADEME Agence de la Transition Écologique. Available online: https://librairie.ademe.fr/energies-renouvelables-reseaux-et-stockage/818-reseaux-de-chaleur-et-de-froid-etat-des-lieux-de-la-filiere-marches-emplois-couts.html (accessed on 26 July 2022).

energies MDPI

Article

Evaluation of the Effectiveness of Different LED Irradiators When Growing Red Mustard (*Brassica juncea* L.) in Indoor Farming

Natalya A. Semenova [1,*], Alexandr A. Smirnov [1], Alexey S. Dorokhov [1], Yuri A. Proshkin [1], Alina S. Ivanitskikh [1], Narek O. Chilingaryan [1], Artem A. Dorokhov [1], Denis V. Yanykin [2], Sergey V. Gudkov [1,2] and Andrey Yu. Izmailov [1]

[1] Federal State Budgetary Scientific Institution "Federal Scientific Agroengineering Center VIM" (FSAC VIM), 109428 Moscow, Russia

[2] Prokhorov General Physics Institute of the Russian Academy of Sciences, Vavilov St., 38, 119991 Moscow, Russia

* Correspondence: natalia.86@inbox.ru; Tel.: +7-926-783-0374

Abstract: Investigation is devoted to the optimization of light spectrum and intensity used for red mustard growing. Notably, most of the studies devoted to red mustard growing were conducted on micro-greens, which is not enough for the development of methods and recommendations for making the right choices about the irradiation parameters for full-cycle cultivation. In this study, we tested four models of LED with different ratios of blue, green red and far red radiation intensity: 12:20:63:5; 15:30:49:6; 30:1:68:1, in two values of photon flux density (PFD)—120 and 180 μmol m^{-2} s^{-1}—to determine the most effective combination for red mustard growing. The study was conducted in a container-type climate chamber, where the red leaf mustard was cultivated in hydroponics. On the 30th day of cultivation, the plant's morphological, biochemical and chlorophyll fluorescence parameters, and reflection coefficients were recorded. The results indicated that the PFD 120 μmol m^{-2} s^{-1} had a worse effect on both mustard leaf biomass accumulation and nitrate concentration (13–30% higher) in the plants. The best lighting option for growing red mustard was the blue–red spectrum, as the most efficient in terms of converting electricity into biomass (77 Wth/g). This light spectrum contributes to plant development with a larger leaf area (60%) and a fresh mass (54%) compared with the control, which has a maximum similarity in spectrum percentage to the sunlight spectrum. The presence of green and far red radiation with the blue–red light spectrum in various proportions at the same level of PFD had a negative effect on plant fresh mass, leaf surface area and photosynthetic activity. The obtained results could be useful for lighting parameters' optimization when growing red mustard in urban farms.

Keywords: red leaf mustard; light-emitting diode; spectral composition of light; productivity; photosynthesis

check for updates

Citation: Semenova, N.A.; Smirnov, A.A.; Dorokhov, A.S.; Proshkin, Y.A.; Ivanitskikh, A.S.; Chilingaryan, N.O.; Dorokhov, A.A.; Yanykin, D.V.; Gudkov, S.V.; Izmailov, A.Y. Evaluation of the Effectiveness of Different LED Irradiators When Growing Red Mustard (*Brassica juncea* L.) in Indoor Farming. *Energies* **2022**, *15*, 8076. https://doi.org/10.3390/en15218076

Academic Editor: David Herak

Received: 4 October 2022
Accepted: 27 October 2022
Published: 31 October 2022

1. Introduction

The leaf mustard (Korean red mustard) is an annual cruciferous plant (Brassicaceae Burnett), which is a valuable food and officinal crop due to its unique composition and low maintenance for cultivation [1,2]. It is a very popular leafy green in Russia, China and India. Due to its beneficial properties, the leaf mustard is recommended for use for the prevention of chronic cardiovascular system diseases, oncology, diabetes and obesity [3–5].

The leaf mustard can be grown both outdoors and indoors under artificial illumination, which expands the geography of this crop cultivation. Lighting is one of the most significant environmental factors affecting the plant's growth and morphology [6]. Among the other cost elements of vegetable production in greenhouses, the cost of lighting can make up 40–80% of the total [7]. Currently, for plant cultivation in urban farms, in lighting systems, the LEDs of various spectral ranges are used due to their high efficiency, and low heat emission in comparison to the gas-discharge lamps. LEDs allow the spectral composition of light to be easily and quickly changed, causing specified reactions of the plants [8–11]. For crop production, some lighting equipment manufacturers produce a special series of LEDs with increased radiation efficiency in the PAR spectrum [12]. In most of the current studies, the red–blue (RB) LEDs are used. The effect of the red/blue light ratio at various emission intensities on the growth of leafy greens is being studied. It is important that the emission spectra of RB LEDs are in good coincident with the absorption spectra of the plant photosynthetic pigments. Additionally, RB LEDs are considered the most energy-efficient LEDs [13,14]. However, studies show that green light (G) is also necessary for plants as it prevents the inhibition of plant generative development [15,16]. In addition, some researchers note that additional illumination of plants with green (G) and far red (FR) lights can increase the fresh mass of the leafy greens [17]. The green light increases the leaf surface area and reduces the specific leaf mass [18–21]. The addition of G to the BR spectrum results in improvement of the plant's nutritional value as the G spectrum maintains a high net photosynthesis rate and photochemical efficiency [22,23]. However, those light sources with G photons making up more than a 50% share of the total photosynthetic photon flux (PPF) cause plant growth inhibition [24]. It was established that additional G irradiation with a 505 nm peak wavelength (cyan) has a significant positive effect on the photosynthetic pigment content [25]. In addition, the positive role of the G light was proven in the balance maintenance between the biomass production and the synthesis of the secondary metabolites involved in plant defensive reactions [26]. It also affects the activity of nitrate assimilation enzymes [27].

When studying 8, 12, 16, 20, and 24 h photoperiods under blue, red, and far red LED illumination modules at a PFD of 300 μmol m^{-2} s^{-1}, it was found that the 8 h photoperiod resulted in the elongation of hypocotyls and an increase in the leaf area and fresh mass of mustard, red pak choi and tatsoi microgreens. The elongation of plants decreased due to the lengthening of the day from 12:00 to 20:00, and the 24 h photoperiod most suppressed the growth process [28].

Several research groups studied the effect of blue and red light combination treatments on microgreens at different values of the photosynthetic photon flux density (PPFD). Jones-Baumgardt and co-authors [29] studied the impact of PPFD values on microgreens of cabbage, arugula and mustard. For all types of microgreens, it was found that anthocyanin concentration was proportional to the PPFD value. In the other research by the same authors, the effect of LED irradiation on the growth indicators and yield of sunflower, cabbage, arugula and mustard microgreens grown in a greenhouse was studied. During the experiment, various PPFD levels were tested in the range from 17 to 304 μmol m^{-2} s^{-1} with a 16 h-long photoperiod. An increase in PPFD value was accompanied by growth of the dry mass, the stability index and the relative chlorophyll content, and by the formation of leaves of a smaller area in cabbage, arugula and mustard [30]. A decrease in PPFD led to an increase in total nitrogen content and a decrease in the total acidity [31,32].

There are studies devoted to the effect of different proportions of red and blue lights (B) at the same level of PPFD on plant development. Ying and co-authors did not find any differences in the content of chlorophyll, carotenoids and nitrates in the microgreens of mustard, arugula and red cabbage at different proportions of B (from 5% to 30%) and R light (from 70% to 95%) [33]. With all types of microgreens, the change in the share of B light did not affect the total content of the extracted chlorophyll, carotenoids or nitrates [33]. Light intensity, in contrast to light quality, affects the total amount of carotenoids in the microgreens and significantly decreased with increased light intensity [34]. It was noted that the accumulation of phenolic compounds in the cabbage and the mustard was more effective at 30% of blue light in the total illumination; at the same time, the concentration of anthocyanins in arugula and red cabbage correlated well with B light proportion [33]. Mustard microgreen cotyledons darkened under a high proportion of blue light [35]. Note that a high proportion of the blue radiation (from 20% to 50%) can lead to inhibition of the leaf plate growth [36,37] and has a positive effect on the accumulation of macro-and microelements [38,39]. It was found that R light and magenta (450 + 650 nm) light compared to white light promoted the accumulation of both total and individual anthocyanins, while B light was found to be the predominant factor in the a non-anthocyanin phenolic accumulation in mustard microgreens [40].

Recent studies have been focused on plant lighting modes optimization including by the addition of different proportions of G and far-red (FR) to the RB at the same PPFD value [17,39,41,42]. The FR addition treatment can inhibit seed germination; so, it is reasonable to use FR treatment after the germination period [43,44]. The addition of G or FR (plus 50 μmol m^{-2} s^{-1}) to the RB light (4:1) increase both the fresh lettuce shoots mass by 20.5% (in the case of G light) and 40.4% (in the case of FR light) and the dry shoot mass by 24.2% and 45.2%, for green and FR light, respectively [45]. The addition of the FR light of high-intensity to the RB light increased the plant height and the cotyledon area of mustard microgreens and did not affect either fresh or dry biomass, while the addition of G light did not have a significant effect [41].

Most recent studies of red mustard cultivation were carried out on microgreens, which is not enough for developing methods and recommendations for leaf maturity growing and choosing the most effective irradiation parameters. The species-specific reaction of plants to the spectral composition of the light, including the plant growth parameters and their biochemical composition, has been mentioned [46–48]. In order to test the hypothesis about the different responses of plants to the spectral composition of irradiation at different levels of PPFD, it is necessary to comprehensively study the effect of the spectral composition of optical radiation on the productivity of mustard leaf with an emphasis on morpho-biochemical parameters and indicators of plant photosynthesis. The study of the irradiation intensity and the spectral composition of light for red mustard growing will allow us to select the optimal lighting modes to obtain maximum yield with minimal energy consumption. The development of technology elements for growing red mustard can be useful for the introduction and dissemination of this rare crop in city farming.

The purpose of this work is to select the optimal lighting option to reduce energy costs when growing red mustard in indoor farming.

2. Materials and Methods

2.1. Red Musturd Variety

The experiments were carried out on the red leaf mustard of the variety "Red Hill" ('Gavrish', Moscow, Russia). This is a cold-resistant and early-ripening variety (the period of leaf ripening is 25–30 days). It is grown both outdoors and in indoor farming. The leaf rosettes can reach 25–35 cm in length and 50–60 g in mass when grown outdoors.

2.2. Cultivation Conditions

The experiment was conducted in three replications from September 2021 to November 2021 in container-type climate chambers for green crops cultivation. The plants were grown in plastic trays on racks equipped with circulated hydroponics (Figure 1). The seeds were sown into pots with a mineral wool substrate. In order to avoid the FR negative effect, the seeds were germinated in the dark. After the cotyledons appeared, the lighting was turned on. When the first leaves appeared, three plants were left in every pot. Eight light-insulated racks were used for the mustard plant cultivation. The cultivation area for each light treatment was equal to 0.68 m^2. The planting density was 50 plants per 1 m^2. The microclimate in the chamber was maintained by an automatic system. The day/night air temperature was 25/22 °C at a relative humidity of 75%. The carbon dioxide concentration was maintained by the ventilation system and corresponded to atmospheric values of 400–450 ppm. No additional CO_2 was used. For nutrient solutions preparation, the Flora Series® (GHE, Fleurance, France) complex of fertilizers for hydroponics was used. The components of the FloraGro, FloraMicro SW, and FloraBloom kits were used in the following ratio: 2.5:2.0:2.5. The electrical conductivity of the nutrient solution was maintained within 1500–1600 μS/cm.

Figure 1. The appearance of mustard plants of 'Red Hill' variety in the phase of first true leaf development under 4 types of lighting with the following spectral composition (B:G:R:FR): Control (15:32:42:11), V1 (12:20:63:5), V2 (15:30:49:6), V3 (30:1:68:1).

2.3. Irradiation Conditions

In the phytochamber, irradiation was provided by combined irradiators based on LEDs of various spectral compositions. As a control, full-spectrum LED luminaires based on Refond RF-W40QI35DS-DF-J-Y LEDs with a 4000 K color temperature and a color rendering index of Ra95 manufactured by Shenzhen Refond Optoelectronics Co., Ltd. (Shenzhen, China) were used. The emission spectrum is as close as possible to the sunlight spectrum. The rest of the irradiators of our own production consisted of combinations of the latest generation LEDs of TM LUXEON 2835 for plant growing: there were blue LEDs (445 nm), red ones (630 nm and 660 nm), far-red ones (730 nm) and white ones with a color temperature of 3000 K. In V1 and V2 modes, 20% and 30% of green light was respectively chosen, since it is known that the addition of G light to B and R can increase shoot mass [15,43]. In the V 3 mode, only red and blue LEDs were used; their percentage was chosen based on the known data, in which anthocyanins and phenolic compound concentrations were at a maximum [33,35]. The LEDs were manufactured by Lumileds Holding B.V. (San Jose, CA, USA). The light period was 16 h. The irradiation parameters are presented in Table 1. Two variants of PFD were used—120 and 180 μmol m^{-2} s^{-1}.The measurements of the photon flux density and the spectral composition of the irradiation were carried out using the MK350D Compact Spectrometer (UPRtek Corp. Miaoli County, Taiwan). For the irradiation variants with 120 μmol m^{-2} s^{-1} PFD, the specific power consumption amounted to 105 $\pm$ 1.5 W m^{-2}. For the lighting system with 180 μmol m^{-2} s^{-1} PFD, the specific power consumption was 140 $\pm$ 2.1 W m^{-2}. The measurements were carried out using a CVM-MINI CIRCUTOR electric power parameter meter (Barcelona, Spain).

Table 1. Average density of photon flux coming from LEDs in each of spectrum zones: Blue (B), Green (G), Red (R) and Far-Red (FR) for s cultivation of "Red Hill" variety red leaf mustard (*Brassica juncea* L.) in the climatic chamber. Average values obtained in five measurement sessions are presented here.

Irradiation Variant	Photon Flux, μmol Photons m^{-2} s^{-1}						Percentage Composition of Light (B:G:R:FR)
	PFD (400 nm–800 nm)	Blue (400 nm–500 nm)	Green (500 nm–600 nm)	Red (600 nm–700 nm)	Far Red (700 nm–800 nm)	PPFD (400 nm–700 nm)	
Control	120 $\pm$ 2.8	17.5 $\pm$ 0.3	38.5 $\pm$ 1.2	51.0 $\pm$ 1.5	13.0 $\pm$ 0.2	107.0 $\pm$ 2.7	15:32:42:11
	180 $\pm$ 3.3	26.4 $\pm$ 0.6	58.0 $\pm$ 1.5	76.0 $\pm$ 1.9	19.6 $\pm$ 0.4	160.4 $\pm$ 3.2	
V1	120 $\pm$ 2.3	14.2 $\pm$ 0.2	26.2 $\pm$ 1.1	73.2 $\pm$ 1.6	6.4 $\pm$ 0.1	113.6 $\pm$ 2.2	12:20:63:5
	180 $\pm$ 3.1	21.4 $\pm$ 0.5	36.7 $\pm$ 1.3	112.9 $\pm$ 1.5	9.0 $\pm$ 0.3	171.0 $\pm$ 3.0	
V2	120 $\pm$ 3.0	18.0 $\pm$ 0.2	36.0 $\pm$ 0.9	58.8 $\pm$ 0.9	7.2 $\pm$ 0.1	112.8 $\pm$ 2.8	15:30:49:6
	180 $\pm$ 3.3	26.2 $\pm$ 0.7	53.9 $\pm$ 1.6	88.8 $\pm$ 1.1	11.3 $\pm$ 0.5	168.9 $\pm$ 3.1	
V3	120 $\pm$ 1.5	36.5 $\pm$ 0.3	1.5 $\pm$ 0.1	81.0 $\pm$ 1.0	1.0 $\pm$ 0.1	119.0 $\pm$ 1.5	30:1:68:1
	180 $\pm$ 3.8	54.5 $\pm$ 0.7	2.0 $\pm$ 0.2	122.0 $\pm$ 2.3	1.5 $\pm$ 0.1	178.5 $\pm$ 3.8	

2.4. Evaluation of Biometric Indicators

The sampling of the mustard plants was carried out according to the principle of representativeness, i.e., in such a way that the characteristics of the selected plants corresponded to that of the entire set of plants of the certain experiment variant. For analyzing the fresh and dry leaf mass indicators and the leaf surface area, we selected 10 plants from each tested variant. The measurements were made on the 30th day after germination.

When determining the fresh mass, the plants were weighed on the laboratory scale GF-3000 (A&D Company, Japan). For the measurement of the leaf surface area, the photo-planimeter LI-COR—LI-3100 AREA METER (LI-COR, Inc. Lincoln, NE, USA) was used. In order to determine dry mass, the samples were crushed to a particle size of no more than 1 cm, dried in an oven at a temperature of 60–70 °C for 3 h until a constant mass was obtained, weighed on the Sartorius LA230S analytical scales (Laboratory Scale, Göttingen, Germany).

2.5. Pigment Content

To determine the content of pigments, the third leaves from the top (fully illuminated) were used. The samples of the fresh materials were crushed in porcelain mortars with the addition of quartz sand. All procedures were repeated three times.

In order to determine the chlorophyll and carotenoid content, we selected samples with 0.1 g mass, and ground them in the mortar with the addition of 2–3 mL of 100% acetone to obtain an extract. Then, the extract was moved to a funnel with a glass filter (No.3) inserted into the Bunsen flask and connected to a vacuum pump. The mortar and the filter were washed repeatedly with 100% acetone until the pigments were completely extracted. Then, the filtrate was transferred to a measuring bottle and the acetone was added up to 25 mL. The contents of the measuring bottles were thoroughly mixed and used for the evaluation of the pigment contents by the spectrometric method on the SPECS SSP-705 spectrophotometer (Russia). The optical density of the pigment solution was determined at wavelengths of 662 nm (for the chlorophyll a), 644 nm (for the chlorophyll b) and 440.5 nm (for the carotenoids). The thickness of the cuvette absorbing layer was 10 mm. The pigment concentration was calculated using the Holm–Wettstein method for 100% acetone [43].

2.6. Nitrates Content

For the nitrate content determination in the leaf mustard samples, we selected three plants in each variant of the experiment. We took samples of 10 g in fresh mass and crushed them in the mortar. The crushed material was placed in a measuring beaker and 50 cm^3 of the extraction solution was topped up. For plants of the Brassicaceae family, this solution was prepared as follows: 1 g of potassium permanganate and 0.6 cm^3 of concentrated sulfuric acid were dissolved in a 1% solution of aluminum–potassium alum, and the volume of the solution was brought up to 1000 cm^3. Then, the material obtained in the measuring beaker was mixed for 3 min in the magnetic stirrer. The concentration of the nitrate ions in the suspension was measured using the ion-meter "Itan" (Tom'analit, Tomsk, Russia).

2.7. Measurement of Chlorophyll Fluorescence Parameters and Vegetation Indices

To measure the activity of the light stage of photosynthesis, a portable fluorimeter, PAR-FluorPen FP 110-LM/D (Photon Systems Instruments, Drásov, Czech Republic), was used to detect active chlorophyll fluorescence and was further analyzed using the PAM method or OJIP test. The PAR-FluorPen FP 110-LM/D consists of a detector (PIN photodiode with a narrow band filter, working optical range from 667 to 750 nm) and a blue LED emitter (maximum about 455 nm), and a sensor of ambient light. To assess the photosynthetic efficiency of photosystem II (Fv/Fm), the leaf was preliminarily dark-adapted for at least 20 min. To determine the spectral reflectance of the leaves, a portable PolyPen RP 410/UVIS meter (Czech Republic) was used. The spectra of nine plants were measured in each lighting variant and at each cultivation period. Three spectral measurements were carried out on different leaves of each individual plant. Using the PolyPen RP 410 UVIS program, the main reflectance indicators and the vegetation index were calculated.

2.8. Energy Intensity Calculation

The cost of electricity for the production of 1 g mustard plant fresh mass was calculated as the ratio of the total electricity consumed by lamps for 30 days per the total mass of all plants under the illuminator of this type.

2.9. Statistical Data Processing

All experiments were carried out threefold. Statistical processing of measurement results and plotting were performed in Python 3.9. To estimate statistical significance, the Independent two-sample T-test was used with $p < 0.05$ significance levels.

3. Results

3.1. Morphology

The measurement results of the morphological parameters of the red leaf mustard plants on the 30th day after their germination are shown in Figure 2. Statistical analysis of the data obtained showed the reliability of differences in terms of fresh mass and leaf surface area. At a PFD of 180 µmol m^{-2} s^{-1}, the fresh mass and leaf area indicators were higher in all experimental variants.

Figure 2. Red leaf mustard (*Brassica juncea* L.) plants of the "Red Hill" variety under 120 µmol m^{-2} s^{-1} PFD (**a**) and 180 µmol m^{-2} s^{-1} PFD (**b**) on 30th day of cultivation.

The best lighting option for fresh mass accumulation and the largest leaf surface area was observed under the spectrum ratio red–blue (V3). The increase in fresh mass was 71.6% (Figure 3a), and in leaf surface area it was 51.8% (Figure 3c). In terms of dry mass (Figure 3b) and number of leaves (the data are not shown), there were no significant differences between the variants in the experiment. Based on the dry mass indicator, the plants of all variants were not inferior to each other in terms of nutritional properties.

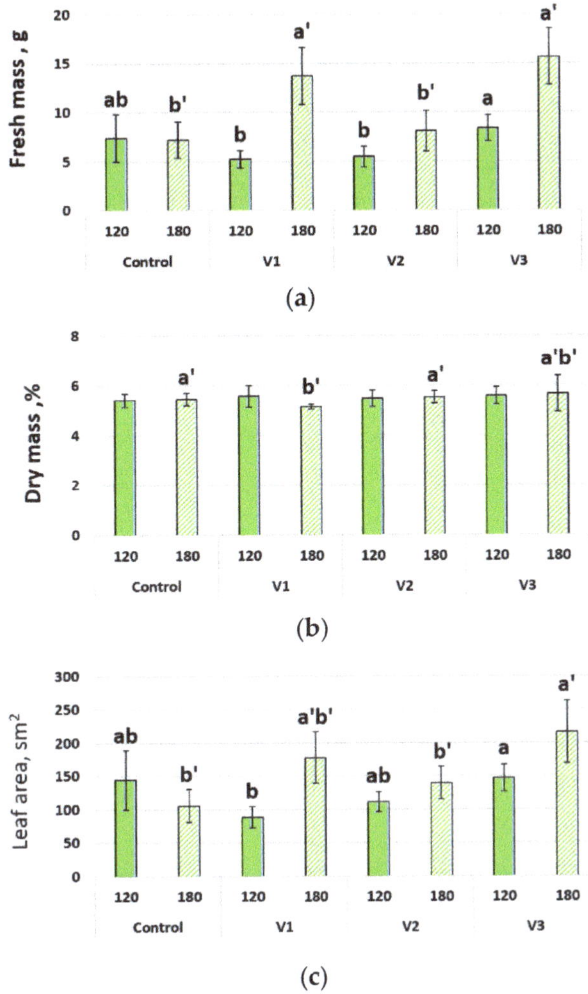

Figure 3. Fresh mass (**a**), dry mass (**b**) and leaf surface area (**c**) of red leaf mustard (*Brassica juncea* L.) plants of the "Red Hill" variety on 30th day of cultivation. The data shown are the means and the vertical bars indicate standard errors (n = 5). Statistically significant differences within $p = 0.05$. The different letters indicate significant differences among groups.

By the 30th day of red mustard cultivation, flowering was not observed in any of the light options. This was facilitated by the optimal length of daylight hours; therefore, in this case, a high proportion of the green part of the spectrum is not required to prevent flowering [49].

In the variant V3, the shares of the far-red and green LEDs are small and plant shadow avoidance reaction [17] was not observed; this is due to the greater photosynthetic activity of red and blue light. The largest leaf surface area was observed in all treatments at PFD 180 μmol m^{-2} s^{-1}; this observation coincides with other studies [30].

3.2. Biochemical Analysis of Leaves

The content of the main photosynthetic pigments is shown in Figure 4. In terms of the carotenoid ratio to total chlorophyll (1:5), we can say that the plants of all variants in the experiment were in the active growth phase and had no signs of aging. In relation to chlorophyll a to chlorophyll b (3:1), it can be concluded that mustard plants of all variants in the experiment did not suffer from a lack of illumination. An increase in the total level of PFD had no significant effect on the concentration of photosynthetic pigments. The best light variations for total chlorophyll content were V3 with light intensity 120 and 180 µmol m^{-2} s^{-1} and V2 with light intensity 120 µmol m^{-2} s^{-1}. In another study conducted on mustard microgreens, total integrated chlorophyll of higher values was observed at the spectral ratio B 13%, R 87% (without addition of G or FR) as compared to B 9%, R 84%, FR 7% and B 8%, G 18%, R74% spectral ratios [34]. It is possible that the negative effect of FR on the total chlorophyll content rises with increasing PPF; therefore in the V2 variant and at 120 µmol m^{-2} s^{-1} PFD, higher chlorophyll content was observed.

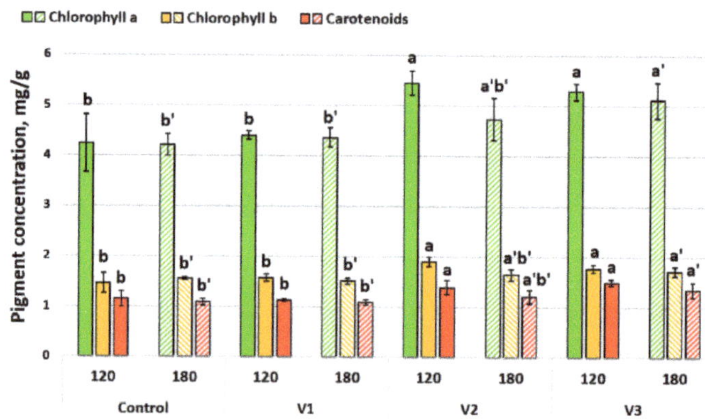

Figure 4. Pigments concentration in red leaf mustard (*Brassica juncea* L.) plants of the "Red Hill" variety on 30th day of cultivation. The data shown are the means and the vertical bars indicate standard errors (n = 5). Statistically significant differences within $p \leq 0.05$. The different letters indicate significant differences among groups.

The nitrate concentration is inversely correlated with the PFD total level (Figure 5). For all variants in the experiment, a PFD of 120 µmol m^{-2} s^{-1} is above the maximum concentration allowable for leafy greens in the Russian Federation [50], so it is unacceptable to use this level of illumination.

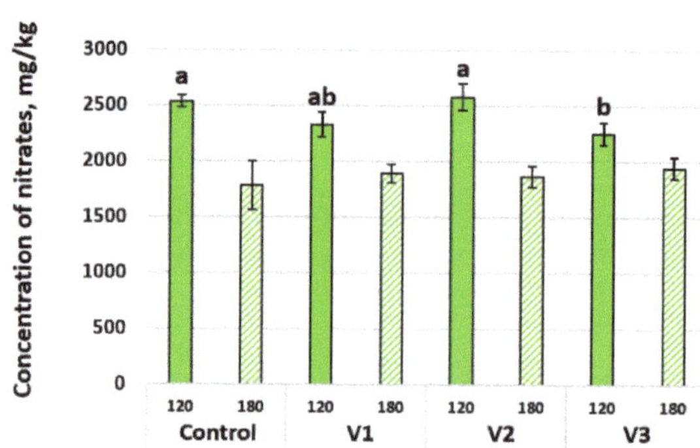

Figure 5. Concentration of nitrates in red leaf mustard (*Brassica juncea* L.) plants of the "Red Hill" variety on 30th day of cultivation. The data shown are the means and the vertical bars indicate standard errors (n = 5). Statistically significant differences within $p \leq 0.05$. The different letters indicate significant differences among groups.

It is known that the combination of red and blue lights with a high share of blue light (25% and higher) can be effective at reducing the nitrate content in the mustard microgreens [38,51]. At the stage before harvesting, the simultaneous increase of PPFD (up to 300 µmol m^{-2} s^{-1}) and blue light share significantly reduces the nitrates concentration in the plants of the Tatsoy cabbage (*Brassica rapa* subsp. *narinosa*) [52]. Additionally, a decreased PPFD value results in the highest nitrate content in the mustard microgreens [53]. In our experimental variants, the share of the G-radiation in the total illumination did not correlate with the nitrate content, in contrast to the studies conducted on lettuce plants [22].

3.3. Chlorophyll Fluorescence Parameters and Measurements of Vegetation Indexes

Figure 6 shows the effect of different lighting options, when growing mustard, on the effective quantum yield (Qy) of photochemical reactions of photosystem II, non-photochemical quenching (NPQ), and photochemical fluorescence quenching coefficient (Qp), which show the state of the plant photosynthesis system. NPQ is a mechanism employed by plants to protect themselves from excessive light intensity. In the V3 variant, an increase in NPQ was observed with an increase in PFD from 120 µmol m^{-2} s^{-1} to 180 µmol m^{-2} s^{-1}, which indicates that at a PFD 120 µmol m^{-2} s^{-1}, the photosynthetic apparatus of the plant was not in saturation, in contrast to other options, where NPQ remained practically unchanged with increasing PFD. On petunia plants, it was found that, under the blue–red spectrum, plant leaves have a lower index Qy [54] than in variants with a white diodes addition, but in our study the blue–red spectrum (V3) had no significant effect on red mustard's Qy index.

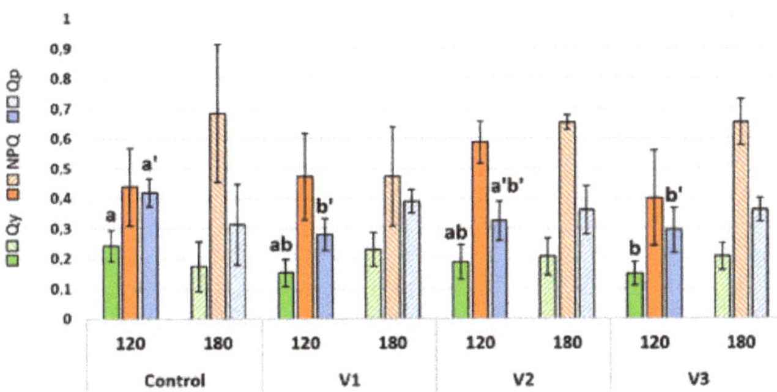

Figure 6. Chlorophyll fluorescence parameters in red mustard plants of the "Red Hill" variety on 30th day of cultivation depending on the type of radiation: Qy—effective quantum yield of photochemical reactions of photosystem II of light-adapted plants, non-photochemical quenching (NPQ), Qp—coefficient of photochemical fluorescence quenching. The different letters indicate significant differences among groups.

The maximal photochemical efficiency of photosystem II ($(F_m-F_o)/F_m$) (data not shown) had no significant differences between the variants and was in the range of 0.83 ÷ 0.84, which means that the photosynthetic apparatus was not damaged in the plants and they did not suffer from light stress [55].

Using a portable spectroradiometer, some vegetation indices were determined (Figure 7). A normalized difference vegetation index (NDVI) serves as an indicator of the plant's state and shows the total amount of green vegetation. PRI is a photochemical reflection index associated with the xanthophyll cycle, which allows us to determine the stress state of the plant's photosynthetic apparatus. (R780-R710)/(R780-R680) is the reflectance index associated with the concentration of chlorophylls [56,57]. The leaf reflection coefficients did not differ significantly from each other in the variants of the experiment, which may be due to slight differences between the variants in the concentrations of pigments (Figure 4), morphological parameters of the plants, and anatomical features of the leaf structure [57].

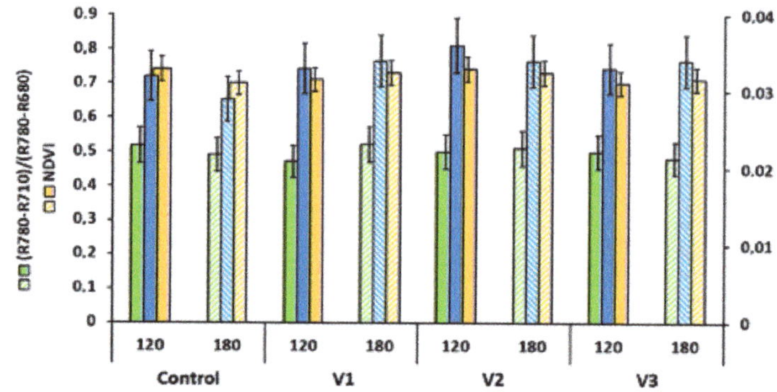

Figure 7. Reflection coefficients of red leaf mustard (*Brassica juncea* L.) plants of the "Red Hill" variety on 30th day of cultivation.

The leaf reflection coefficients did not differ significantly from each other in the variants of the experiment, which may be due to slight differences between the variants in the concentrations of pigments (Figure 4).

3.4. Energy Intensity

The electricity costs for the production of 1 g of mustard leaf at 30 days of cultivation were calculated (Figure 8). The lowest value of consumed energy was in variant V3 with a PFD of 180 µmol m^{-2} s^{-1} (77 Wt h/g). The highest value was observed in the V1 variant at a PFD of 120 µmol m^{-2} s^{-1} (190 Wt h/g). It can be associated with the efficiency of plant photon absorption of different spectral ranges. In the V3 variant, the irradiators contained mainly red and blue LEDs, which have the highest photon output of any LED available [58]. In addition, blue and red light are well absorbed by chlorophylls and carotenoids. The V3 variant was distinguished by a higher concentration of these photosynthetic pigments.

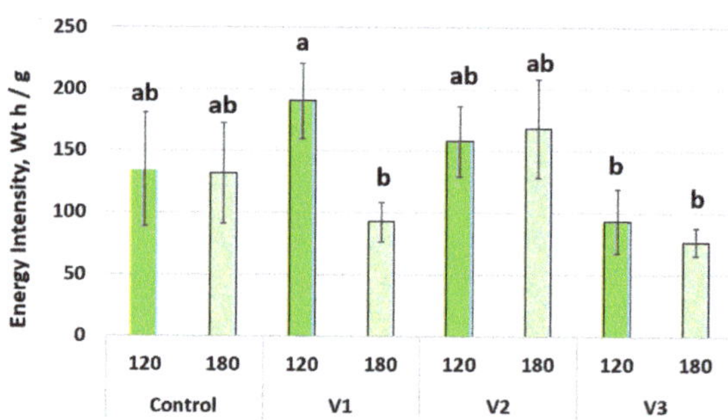

Figure 8. Energy intensity of cultivation of red leaf mustard (*Brassica juncea* L.) plants of the "Red Hill" variety on 30th day. The data shown are the means and the vertical bars indicate standard errors ($n = 5$). Statistically significant differences within $p \leq 0.05$. The different letters indicate significant differences among groups.

4. Discussion

Plants mainly absorb blue light and red light, which indicates the possibility of growing with red and blue LEDs [59]. Therefore, lettuce [60–62], radishes [63] and tomatoes [64] have been studied by numerous researchers using a combination of red and blue light. In this study, under red–blue LEDs, mustard leaf had a larger leaf area and fresh weight by 50% and 70%, respectively. Control and V2 plants had the lowest leaf area and fresh weight due to a higher green light PFD percentage (30%). In addition to red and blue LEDs, green and far red diodes were added to the growth chambers, as a combination of blue, red, green and far red light in small amounts is most suitable for crop growth [65]. Similar systems, for example, are installed on the International Space Station [66], at the Chinese space laboratory, Tiangong II [63], at the Vegetable Production System [66–71] and in the Advanced Plant Habitat [72].

It is well known that high and low light intensities, regardless of the spectral composition, inhibit the efficiency of photosynthesis, in particular the effective quantum yield [71]. In our work, we used different light intensities of 120 and 180 μmol photons m^{-2} s^{-1}; 120 μmol photons m^{-2} s^{-1} was not enough to completely saturate the photosynthetic apparatus of plants, as evidenced by an increase in NPQ and a decrease in Qy in leaves. It was also shown that in all experimental cases there was a decrease in NPQ and Qp. This is due to the presence of green and far red light in the spectrum, which stabilizes the photosynthetic apparatus and reduces the amount of photodamage in photosystem II [72]. Furthermore, the addition of a small amount of green light and far red light has been shown to save plants from photoinhibition when exposed to prolonged light [73,74]. These results show that our LED system is able to improve the growth and development of mustard.

5. Conclusions

When growing red mustard of the 'Red Hill" variety, the best lighting option was the red–blue (V3) with an irradiation intensity of 180 μmol m^{-2} s^{-1}. This intensity meets the needs of mustard plants, and the spectral composition, with the highest proportion of blue and red illumination among the selected options, contributes to the formation of compact dense rosettes of leaves with a larger leaf area and fresh mass. Additional illumination of plants with green and far-red light in variants with the addition of white LEDs for mustard did not contribute to an increase in fresh mass, in contrast to other studies of green crops [15]. The addition of green and far red radiation to blue–red light in various proportions at the same level of PFD, in contrast to studies on microgreens [39], had a negative effect on plant fresh mass, leaf surface area and photosynthetic activity. In future studies we are going to identify the optimal proportion of blue and red light and variety-specific reactions when growing red mustard of various varieties.

The results obtained will be useful for the optimization of the artificial lighting parameters of red mustard cultivation in urban farms.

Author Contributions: Conceptualization and methodology, N.O.C. and D.V.Y.; validation, N.A.S. and A.S.I.; formal analysis, N.A.S. and A.A.D.; investigation and resources, A.A.S. and N.O.C.; writing—original draft preparation, A.A.S., D.V.Y. and N.A.S.; writing—review and editing, N.A.S., A.A.S., A.S.I. and S.V.G., visualization, Y.A.P.; supervision and funding acquisition, A.Y.I.; project administration, A.S.D. All authors have read and agreed to the published version of the manuscript.

Funding: This research was funded by a grant of the Ministry of Science and Higher Education of the Russian Federation for large scientific projects in priority areas of scientific and technological development (subsidy identifier 075-15-2020-774).

Institutional Review Board Statement: Not applicable.

Informed Consent Statement: Not applicable.

Data Availability Statement: No additional data available.

Conflicts of Interest: The authors declare no conflict of interest. The funders had no role in the design of the study, in the collection, analyses, or interpretation of the data, in the writing of the manuscript, or in the decision to publish the results.

References

1. Kim, Y.T.; Kim, B.K.; Park, K.Y. Antimutagenic and anticancer effects of leaf mustard and leaf mustard kimchi. *J. Korean Soc. Food Sci. Nutr.* **2007**, *12*, 84–88. [CrossRef]
2. Tian, Y.; Deng, F. Phytochemistry and biological activity of mustard (*Brassica juncea*): A review. *CyTA-J. Food* **2020**, *18*, 704–718. [CrossRef]

3. Sturm, C.; Wagner, A.E. Brassica-Derived Plant Bioactives as Modulators of Chemopreventive and Inflammatory Signaling Pathways. *Int. J. Mol. Sci.* **2017**, *18*, 1890. [CrossRef]
4. Raiola, A.; Errico, A.; Petruk, G.; Monti, D.M.; Barone, A.; Rigano, M.M. Bioactive Compounds in Brassicaceae Vegetables with a Role in the Prevention of Chronic Diseases. *Molecules* **2017**, *23*, 15. [CrossRef]
5. Jo, S.-H.; Cho, C.-Y.; Ha, K.-S.; Lee, J.-Y.; Choi, H.-Y.; Kwon, Y.-I.; Apostolidis, E. In vitro and in vivo anti-hyperglycemic effects of green and red mustard leaves (*Brassica juncea* var. integrifolia). *J. Food Biochem.* **2018**, *42*, e12583. [CrossRef]
6. Naznin, M.T.; Lefsrud, M. An Overview of LED Lighting and Spectral Quality on Plant Photosynthesis. In *Light Emitting Diodes for Agriculture*; Dutta Gupta, S., Ed.; Springer: Singapore, 2017. [CrossRef]
7. Katzin, D.; Marcelis, L.F.; van Mourik, S. Energy savings in greenhouses by transition from high-pressure sodium to LED lighting. *Appl. Energy* **2020**, *281*, 116019. [CrossRef]
8. Runkle, E.; Meng, Q.; Park, Y. LED applications in greenhouse and indoor production of horticultural crops. *Acta Hortic.* **2019**, *1263*, 17–30. [CrossRef]
9. Paradiso, R.; Proietti, S. Light-Quality Manipulation to Control Plant Growth and Photomorphogenesis in Greenhouse Horticulture: The State of the Art and the Opportunities of Modern LED Systems. *J. Plant Growth Regul.* **2021**, *41*, 742–780. [CrossRef]
10. Gudkov, S.V.; Andreev, S.N.; Barmina, E.V.; Bunkin, N.F.; Kartabaeva, B.B.; Nesvat, A.P.; Stepanov, E.V.; Taranda, N.I.; Khramov, R.N.; Glinushkin, A.P. Effect of visible light on biological objects: Physiological and pathophysiological aspects. *Phys. Wave Phenom.* **2017**, *25*, 207–213. [CrossRef]
11. Paskhin, M.O.; Yanykin, D.V.; Gudkov, S.V. Current Approaches to Light Conversion for Controlled Environment Agricultural Applications: A Review. *Horticulturae* **2022**, *8*, 885. [CrossRef]
12. Nair, G.B.; Dhoble, S.J. *The Fundamentals and Applications of Light-Emitting Diodes*; Nair, G.B., Ed.; Woodhead Publishing: Soston, UK, 2021; Chapter 5; pp. 127–152.
13. Mitchell, C.A.; Dzakovich, M.P.; Gomez, C.; Lopez, R.; Burr, J.F.; Hernández, R.; Kubota, C.; Currey, C.J.; Meng, Q.; Runkle, E.S.; et al. Light-Emitting Diodes in Horticulture. *Hortic. Rev.* **2015**, *43*, 1–88. [CrossRef]
14. Mengxi, L.; Zhigang, X.; Yang, Y.; Yijie, F. Effects of different spectral lights on Oncidium PLBs induction, proliferation, and plant regeneration. *Plant Cell Tissue Organ Cult.* **2010**, *106*, 1–10. [CrossRef]
15. Avercheva, O.; Berkovich, Y.A.; Smolyanina, S.; Bassarskaya, E.; Pogosyan, S.; Ptushenko, V.; Erokhin, A.; Zhigalova, T. Biochemical, photosynthetic and productive parameters of Chinese cabbage grown under blue–red LED assembly designed for space agriculture. *Adv. Space Res.* **2014**, *53*, 1574–1581. [CrossRef]
16. Kasajima, S.-Y.; Inoue, N.; Mahmud, R.; Kato, M. Developmental Responses of Wheat cv. Norin 61 to Fluence Rate of Green Light. *Plant Prod. Sci.* **2008**, *11*, 76–81. [CrossRef]
17. Meng, Q.; Kelly, N.; Runkle, E.S. Substituting green or far-red radiation for blue radiation induces shade avoidance and promotes growth in lettuce and kale. *Environ. Exp. Bot.* **2019**, *162*, 383–391. [CrossRef]
18. Claypool, N.; Lieth, J. Physiological responses of pepper seedlings to various ratios of blue, green, and red light using LED lamps. *Sci. Hortic.* **2020**, *268*, 109371. [CrossRef]
19. Kim, H.-H.; Wheeler, R.; Sager, J.; Goins, G. A Comparison of Growth and Photosynthetic Characteristics of Lettuce Grown under Red and Blue Light-Emitting Diodes (LEDS) with and without Supplemental Green LEDS. *Acta Hortic.* **2004**, *659*, 467–475. [CrossRef]
20. Kamal, K.Y.; Khodaeiaminjan, M.; El-Tantawy, A.A.; El Moneim, D.A.; Salam, A.A.; Ash-Shormillesy, S.M.A.I.; Attia, A.; Ali, M.A.S.; Herranz, R.; El-Esawi, M.A.; et al. Evaluation of growth and nutritional value of Brassica microgreens grown under red, blue and green LEDs combinations. *Physiol. Plant.* **2020**, *169*, 625–638. [CrossRef]
21. Li, L.; Tong, Y.-X.; Lu, J.-L.; Li, Y.-M.; Yang, Q.-C. Lettuce Growth, Nutritional Quality, and Energy Use Efficiency as Affected by Red–Blue Light Combined with Different Monochromatic Wavelengths. *HortScience* **2020**, *55*, 613–620. [CrossRef]
22. Bian, Z.; Cheng, R.; Wang, Y.; Yang, Q.; Lu, C. Effect of green light on nitrate reduction and edible quality of hydroponically grown lettuce (*Lactuca sativa* L.) under short-term continuous light from red and blue light-emitting diodes. *Environ. Exp. Bot.* **2018**, *153*, 63–71. [CrossRef]
23. Lim, S.; Kim, J. Light Quality Affects Water Use of Sweet Basil by Changing Its Stomatal Development. *Agronomy* **2021**, *11*, 303. [CrossRef]
24. Kim, H.; Wheeler, R.; Sager, J.; Gains, G.; Naikane, J. Evaluation of Lettuce Growth Using Supplemental Green Light with Red and Blue Light-Emitting Diodes in a Controlled Environment—A Review of Research at Kennedy Space Center. *Acta Hortic.* **2006**, *711*, 111–120. [CrossRef]
25. Sirtautas, R.; Viršilė, A.; Samuolienė, G.; Brazaitytė, A.; Miliauskienė, J.; Sakalauskienė, S.; Duchovskis, P. Growing of leaf lettuce (*Lactuca sativa* L.) under high-pressure sodium lamps with supplemental blue, cyan and green LEDs. *Zemdirb. Agric.* **2014**, *101*, 75–78. [CrossRef]
26. Kitazaki, K.; Fukushima, A.; Nakabayashi, R.; Okazaki, Y.; Kobayashi, M.; Mori, T.; Nishizawa, T.; Reyes-Chin-Wo, S.; Michelmore, R.W.; Saito, K.; et al. Metabolic Reprogramming in Leaf Lettuce Grown Under Different Light Quality and Intensity Conditions Using Narrow-Band LEDs. *Sci. Rep.* **2018**, *8*, 7914. [CrossRef]
27. Yang, P.; Wang, Y.; Li, J.; Bian, Z. Effects of Brassinosteroids on Photosynthetic Performance and Nitrogen Metabolism in Pepper Seedlings under Chilling Stress. *Agronomy* **2019**, *9*, 839. [CrossRef]

28. Vaštakaitė-Kairienė, V.; Brazaitytė, A.; Samuolienė, G.; Viršilė, A.; Miliauskienė, J.; Jankauskienė, J.; Novičkovas, A.; Duchovskis, P. The influence of LED light photoperiod on growth and mineral composition of *Brassica* microgreens indoors. *Acta Hortic.* **2022**, *1337*, 143–150. [CrossRef]
29. Jones-Baumgardt, C.; Ying, Q.; Zheng, Y.; Bozzo, G.G. The growth and morphology of microgreens is associated with modified ascorbate and anthocyanin profiles in response to the intensity of sole-source light-emitting diodes. *Can. J. Plant Sci.* **2021**, *101*, 212–228. [CrossRef]
30. Jones-Baumgardt, C.; Llewellyn, D.; Zheng, Y. Different Microgreen Genotypes Have Unique Growth and Yield Responses to Intensity of Supplemental PAR from Light-emitting Diodes during Winter Greenhouse Production in Southern Ontario, Canada. *HortScience* **2020**, *55*, 156–163. [CrossRef]
31. Craver, J.K.; Gerovac, J.R.; Lopez, R.G.; Kopsell, D.A. Light Intensity and Light Quality from Sole-source Light-emitting Diodes Impact Phytochemical Concentrations within Brassica Microgreens. *J. Am. Soc. Hortic. Sci.* **2017**, *142*, 3–12. [CrossRef]
32. Makus, J.; Zibilske, L.; Lester, G. Effect of light intensity, soil type, and lithium addition on spinach and mustard greens leaf constituents. *Subtrop. Plant Sci.* **2006**, *58*, 35.
33. Ying, Q.; Jones-Baumgardt, C.; Zheng, Y.; Bozzo, G. The Proportion of Blue Light from Light-emitting Diodes Alters Microgreen Phytochemical Profiles in a Species-specific Manner. *HortScience* **2021**, *56*, 13–20. [CrossRef]
34. Samuolienė, G.; Viršilė, A.; Brazaitytė, A.; Jankauskienė, J.; Sakalauskienė, S.; Vaštakaitė, V.; Novičkovas, A.; Viškelienė, A.; Sasnauskas, A.; Duchovskis, P. Blue light dosage affects carotenoids and tocopherols in microgreens. *Food Chem.* **2017**, *228*, 50–56. [CrossRef] [PubMed]
35. Ying, Q.; Kong, Y.; Jones-Baumgardt, C.; Zheng, Y. Responses of yield and appearance quality of four Brassicaceae microgreens to varied blue light proportion in red and blue light-emitting diodes lighting. *Sci. Hortic.* **2020**, *259*, 108857. [CrossRef]
36. Yan, Z.; He, D.; Niu, G.; Zhou, Q.; Qu, Y. Growth, Nutritional Quality, and Energy Use Efficiency of Hydroponic Lettuce as Influenced by Daily Light Integrals Exposed to White versus White Plus Red Light-emitting Diodes. *HortScience* **2019**, *54*, 1737–1744. [CrossRef]
37. Meng, Q.; Runkle, E.S. Growth Responses of Red-Leaf Lettuce to Temporal Spectral Changes. *Front. Plant Sci.* **2020**, *11*, 571788. [CrossRef] [PubMed]
38. Brazaitytė, A.; Miliauskienė, J.; Vaštakaitė-Kairienė, V.; Sutulienė, R.; Laužikė, K.; Duchovskis, P.; Małek, S. Effect of Different Ratios of Blue and Red LED Light on Brassicaceae Microgreens under a Controlled Environment. *Plants* **2021**, *10*, 801. [CrossRef]
39. Meng, Q.; Boldt, J.; Runkle, E.S. Blue Radiation Interacts with Green Radiation to Influence Growth and Predominantly Controls Quality Attributes of Lettuce. *J. Am. Soc. Hortic. Sci.* **2020**, *145*, 75–87. [CrossRef]
40. Liu, Z.; Teng, Z.; Pearlstein, D.J.; Chen, P.; Yu, L.; Zhou, B.; Luo, Y.; Sun, J. Effects of Different Light-Emitting Diode Illuminations on Bioactive Compounds in Ruby Streaks Mustard Microgreens by Ultra-High Performance Liquid Chromatography–High-Resolution Mass Spectrometry. *ACS Food Sci. Technol.* **2022**, *2*, 9. [CrossRef]
41. Ying, Q.; Kong, Y.; Zheng, Y. Growth and Appearance Quality of Four Microgreen Species under Light-emitting Diode Lights with Different Spectral Combinations. *HortScience* **2020**, *55*, 1399–1405. [CrossRef]
42. Yudina, L.; Sukhova, E.; Mudrilov, M.; Nerush, V.; Pecherina, A.; Smirnov, A.A.; Dorokhov, A.S.; Chilingaryan, N.O.; Vodeneev, V.; Sukhov, V. Ratio of Intensities of Blue and Red Light at Cultivation Influences Photosynthetic Light Reactions, Respiration, Growth, and Reflectance Indices in Lettuce. *Biology* **2022**, *11*, 60. [CrossRef]
43. Semenova, N.; Smirnov, A.; Grishin, A.; Pishchalnikov, R.; Chesalin, D.; Gudkov, S.; Chilingaryan, N.; Skorokhodova, A.; Dorokhov, A.; Izmailov, A. The Effect of Plant Growth Compensation by Adding Silicon-Containing Fertilizer under Light Stress Conditions. *Plants* **2021**, *10*, 1287. [CrossRef] [PubMed]
44. Bochenek, G.; Fällström, I. How green is white light? A comparison of basil growth under green or white enriched LED light regimes. *Acta Hortic.* **2015**, *1107*, 311–316. [CrossRef]
45. Shichijo, C.; Katada, K.; Tanaka, O.; Hashimoto, T. Phytochrome A-mediated inhibition of seed germination in tomato. *Planta* **2001**, *213*, 764–769. [CrossRef] [PubMed]
46. Vastakaite, V.; Virsile, A. Light—Emitting Diodes (LEDs) for Higher Nutritional Quality of Brassicaceae Microgreens. *Res. Rural. Dev.* **2015**, *1*, 111–117.
47. Massa, G.; Graham, T.; Haire, T.; Flemming, C.; Newsham, G.; Wheeler, R. Light-emitting Diode Light Transmission through Leaf Tissue of Seven Different Crops. *HortScience* **2015**, *50*, 501–506. [CrossRef]
48. Li, Q.; Deng, M.; Xiong, Y.; Coombes, A.; Zhao, W. Morphological and Photosynthetic Response to High and Low Irradiance of *Aeschynanthus longicaulis*. *Sci. World J.* **2014**, *2014*, 347461. [CrossRef]
49. Nájera, C.; Urrestarazu, M. Effect of the Intensity and Spectral Quality of LED Light on Yield and Nitrate Accumulation in Vegetables. *HortScience* **2019**, *54*, 1745–1750. [CrossRef]
50. Technical Regulation of the Customs Union "On Food Safety" (TR CU 021/2011). Available online: https://eacgroupcompany.com/en/regulations/trcu021-2011 (accessed on 25 September 2022).
51. Brazaitytė, A.; Vaštakaitė-Kairienė, V.; Jankauskienė, J.; Viršilė, A.; Samuolienė, G.; Sakalauskienė, S.; Novičkovas, A.; Miliauskienė, J.; Duchovskis, P. Effect of blue light percentage on mineral elements content in *Brassica* microgreens. *Acta Hortic.* **2020**, *1271*, 119–126. [CrossRef]
52. Simanavicius, L.; Viršilė, A. The effects of led lighting on nitrates, nitrites and organic acids in tatsoi. *Res. Rural. Dev.* **2018**, *2*, 95–99. [CrossRef]

53. Samuolienė, G.; Brazaitytė, A.; Jankauskienė, J.; Viršilė, A.; Sirtautas, R.; Novičkovas, A.; Sakalauskienė, S.; Sakalauskaitė, J.; Duchovskis, P. LED irradiance level affects growth and nutritional quality of Brassica microgreens. *Cent. Eur. J. Biol.* **2013**, *8*, 1241–1249. [CrossRef]
54. Phansurin, W.; Jamaree, T.; Sakhonwasee, S. Comparison of Growth, Development, and Photosynthesis of Petunia Grown Under White or Red-blue LED lights. *Korean J. Hortic. Sci.* **2017**, *35*, 689–699. [CrossRef]
55. Maxwell, K.; Johnson, G.N. Chlorophyll fluorescence—A practical guide. *J. Exp. Bot.* **2000**, *51*, 659–668. [CrossRef] [PubMed]
56. Proshkin, Y.A.; Smirnov, A.A.; Semenova, N.A.; Dorokhov, A.S.; Burynin, D.A.; Ivanitskikh, A.S.; Panchenko, V.A. Assessment of Ultraviolet Impact on Main Pigment Content in Purple Basil (*Ocimum basilicum* L.) by the Spectrometric Method and Hyperspectral Images Analysis. *Appl. Sci.* **2021**, *11*, 8804. [CrossRef]
57. Kior, A.; Sukhov, V.; Sukhova, E. Application of Reflectance Indices for Remote Sensing of Plants and Revealing Actions of Stressors. *Photonics* **2021**, *8*, 582. [CrossRef]
58. Kusuma, P.; Pattison, P.M.; Bugbee, B. From physics to fixtures to food: Current and potential LED efficacy. *Hortic. Res.* **2020**, *7*, 56. [CrossRef]
59. Tang, Y.K.; Guo, S.S.; Ai, W.D.; Qin, L.F. Effects of red and blue light emitting diodes (LEDs) on the growth and devel-opment of lettuce (var. Youmaicai). *Search Technol. Pap.* **2009**, *1*, 2565.
60. Chang, C.-L.; Chang, K.-P. The growth response of leaf lettuce at different stages to multiple wavelength-band light-emitting diode lighting. *Sci. Hortic.* **2014**, *179*, 78–84. [CrossRef]
61. Martineau, V.; Lefsrud, M.; Naznin, M.T.; Kopsell, D. Comparison of Light-emitting Diode and High-pressure Sodium Light Treatments for Hydroponics Growth of Boston Lettuce. *HortScience* **2012**, *47*, 477–482. [CrossRef]
62. Chen, X.-L.; Xue, X.-Z.; Guo, W.-Z.; Wang, L.-C.; Qiao, X.-J. Growth and nutritional properties of lettuce affected by mixed irradiation of white and supplemental light provided by light-emitting diode. *Sci. Hortic.* **2016**, *200*, 111–118. [CrossRef]
63. Samuolienė, G.; Sirtautas, R.; Brazaitytė, A.; Sakalauskaite, J.; Sakalauskienė, S.; Duchovskis, P. The Impact of Red and Blue Light-Emitting Diode Illumination on Radish Physiological Indices. *Cent. Eur. J. Biol.* **2011**, *6*, 821–828. [CrossRef]
64. Naznin, M.T.; Lefsrud, M.G. Impact of LED irradiance on plant photosynthesis and action spectrum of plantlet. In Proceedings of the Optics and Photonics for Information Processing VIII, San Diego, CA, USA, 19 September 2014; p. 921602. [CrossRef]
65. Cope, K.R.; Bugbee, B. Spectral Effects of Three Types of White Light-emitting Diodes on Plant Growth and Development: Absolute versus Relative Amounts of Blue Light. *HortScience* **2013**, *48*, 504–509. [CrossRef]
66. Zabel, P.; Bamsey, M.; Schubert, D.; Tajmar, M. Review and analysis of over 40 years of space plant growth systems. *Life Sci. Space Res.* **2016**, *10*, 1–16. [CrossRef] [PubMed]
67. Shen, Y.; Guo, S.; Zhao, P.; Wang, L.; Wang, X.; Li, J.; Bian, Q. Research on lettuce growth technology onboard Chinese Tiangong II Spacelab. *Acta Astronaut.* **2018**, *144*, 97–102. [CrossRef]
68. Massa, G.; Wheeler, R.; Morrow, R.; Levine, H. Growth chambers on the International Space Station for large plants. *Acta Hortic.* **2016**, *1134*, 215–222. [CrossRef]
69. Bamsey, M.; Graham, T.; Thompson, C.; Berinstain, A.; Scott, A.; Dixon, M. Ion-Specific Nutrient Management in Closed Systems: The Necessity for Ion-Selective Sensors in Terrestrial and Space-Based Agriculture and Water Management Systems. *Sensors* **2012**, *12*, 13349–13392. [CrossRef]
70. Morrow, R.; Richter, R.; Tellez, G.; Monje, O.; Wheeler, R.; Massa, G.; Onate, B. A New Plant Habitat Facility for the ISS. In Proceedings of the 46th International Conference on Environmental Systems, Vienna, Austria, 10–14 July 2016.
71. Ilieva, I.; Ivanova, T.; Naydenov, Y.; Dandolov, I.; Stefanov, D. Plant experiments with light-emitting diode module in Svet space greenhouse. *Adv. Space Res.* **2010**, *46*, 840–845. [CrossRef]
72. Bian, Z.; Yang, Q.; Li, T.; Cheng, R.; Barnett, Y.; Lu, C. Study of the beneficial effects of green light on lettuce grown under short-term continuous red and blue light-emitting diodes. *Physiol. Plant.* **2018**, *164*, 226–240. [CrossRef]
73. Globig, S.; Rosen, I.; Janes, H.W. Continuous Light Effects on Photosynthesis and Carbon Metabolism in Tomato. In *III Inter-National Symposium on Artificial Lighting in Horticulture*; 1994; Volume 418, pp. 141–152. Available online: https://www.actahort.org/books/418 (accessed on 25 September 2022).
74. Demers, D.A.; Gosselin, A. Growing Greenhouse Tomato and Sweet Pepper under Supplemental Lighting: Optimal Photo-Period, Negative Effects of Long Photoperiod and Their Causes. In *IV International ISHS Symposium on Artificial Lighting*; 2000; Volume 580, pp. 83–88. Available online: https://www.actahort.org/books/580 (accessed on 25 September 2022).

energies

Article

Evaluation of Greenhouse Gas Emission Levels during the Combustion of Selected Types of Agricultural Biomass

Jacek Wasilewski, Grzegorz Zając *, Joanna Szyszlak-Bargłowicz and Andrzej Kuranc

Department of Power Engineering and Transportation, Faculty of Production Engineering, University of Life Sciences in Lublin, Głęboka 28, 20-612 Lublin, Poland
* Correspondence: grzegorz.zajac@up.lublin.pl

Abstract: This paper presents the results of an experimental study of the emission levels of selected greenhouse gases (CO_2, CH_4, NO_x) arising from the combustion of different forms of biomass, i.e., solid biomass in the form of pellets and liquid biomass in the example of engine biofuel (biodiesel). Both types of biomass under study are rape-based biofuels. The pellets are made from rape straw, which, as a waste product, can be used for energy purposes. Additionally, biodiesel contains rape oil methyl esters (FAME) designed to power diesel engines. The boiler 25 kW was used to burn the pellets. Engine measurements were performed on a dynamometer bench on an S-4003 tractor engine. An analyzer Testo 350 was used to analyze the exhaust gas. CO_2 emission studies do not indicate the environmental benefits of using any alternative fuels tested compared to their conventional counterparts. In both the engine and boiler tests for NO_x emissions, no environmental benefits were demonstrated from the use of alternative fuels. The measured average NO_x emission levels for biodiesel compared to diesel were about 20% higher, and for rapeseed straw pellets, they were more than 60% higher compared to wood pellets. Only in the case of engine tests was significantly lower CH_4 (approx. 30%) emission found when feeding the engine with rape oil methyl esters.

Keywords: greenhouse gases (GHG); rape straw pellets; biodiesel; combustion; CO_2; CH_4; NO_x emission levels; pellet boiler; tractor engine

Citation: Wasilewski, J.; Zając, G.; Szyszlak-Bargłowicz, J.; Kuranc, A. Evaluation of Greenhouse Gas Emission Levels during the Combustion of Selected Types of Agricultural Biomass. *Energies* **2022**, *15*, 7335. https://doi.org/10.3390/en15197335

Academic Editors: Vadim Bolshev, Vladimir Panchenko, Nallapaneni Manoj Kumar, Pandian Vasant, Igor S. Litvinchev and Prasun Chakrabarti

Received: 14 September 2022
Accepted: 2 October 2022
Published: 6 October 2022

Publisher's Note: MDPI stays neutral with regard to jurisdictional claims in published maps and institutional affiliations.

1. Introduction

The most serious global threats to the natural environment include the intensification of so-called greenhouse gas (GHG) emissions. The effect of this phenomenon's intensification is climatic changes of global character, caused by an increase in the temperature of the lower layers of the atmosphere as well as the surface of the Earth and surface waters [1].

Fossil fuel consumption is a major cause of climate change. In China, where coal-fired power plants dominate, the carbon emission factor is about 1.1 kg $CO_2 \cdot kWh^{-1}$. In the Tokyo area of Japan, the carbon emission factor is about 0.4 kg $CO_2 \cdot kWh^{-1}$, and in regions using hydropower, such as Brazil, it is 0.2 kg $CO_2 \cdot kWh^{-1}$ [2]. NO_x emission values are higher when burning coal fuels compared to other fuels, while CO_2 emissions are the highest when burning lignite [3]. A review of the environmental impact of electricity generation based on combustion technologies of different fuels [4] clearly indicates that hard coal combustion has the highest impact on global warming and ecotoxicity. Among fossil fuels, the highest CO_2 emission factors are characterized by lignite combustion (1300 $kg \cdot MWh^{-1}$), and the lowest are characterized by natural gas (550 $kg \cdot MWh^{-1}$). In contrast, the highest NO_x emissions are associated with the combustion of diesel fuel (75,000 $kg \cdot MWh^{-1}$), and the lowest are associated with natural gas and lignite (15,000 $kg \cdot MWh^{-1}$). For CH_4 emissions, the high emissions are associated with the combustion of hard coal and natural gas (15,000–20,000 $kg \cdot MWh^{-1}$).

Interest in bioenergy has increased in recent decades due to the increased awareness of climate change issues and ambitions to reduce the dependence on fossil fuels [5]. The use of biomass is one of the options for an emission-neutral greenhouse gas as an energy source. However, as [6] noted, carbon neutrality in terms of carbon emissions is not the same as climate neutrality.

Biofuels can reduce the use of fossil fuels and thus reduce greenhouse gas emissions. The impact of biofuel use on mitigating climate change and reducing dependence on fossil fuels is the subject of intense debate in the scientific community. The actual benefits may be limited by local geographic factors, biofuel production technology, and the energy system used.

The use of biomass for energy production carries the risk of increasing emissions of GHG into the atmosphere. According to [7], biomass combustion, including residential biomass combustion (RBC), is a significant global source of gaseous emissions. Biomass combustion is the third largest source of CH_4, contributing to global methane emissions of approximately 40 Tg per year [8] and thus having a direct impact on the global CH_4 balance due to its long residence in the troposphere. Biomass combustion also emits nitrous oxide N_2O. Reactive nitrogen compounds have a significant impact on the chemistry of the atmosphere. This gas has a higher ability to absorb and remit terrestrial radiation than CO_2 or CH_4, so it is a more potent greenhouse gas.

Another use of biomass is biofuels for transportation. Vegetable oils or animal fats are converted to biodiesel through the transesterification with methanol, resulting in a mixture of fatty acid methyl esters (FAME) [9], which are used as an alternative fuel for feeding internal combustion engines [10]. Although the new guidelines mandate that biodiesel be derived from non-food raw materials, for Central Europe and countries with similar climatic conditions, the most promising is the use of biodiesel based on rape oil [11]. In the process of obtaining rapeseed oil, there is waste generated in the form of rapeseed straw, which can be utilized for energy purposes. Some researches [12,13] found out that rape straw is a valuable energy raw material of high calorific value. Redundant rape straw can become raw material for the production of compact solid biofuels [14] as an alternative to wood pellets used in low-power boilers with automatic loading. This policy is in line with one of the objectives of the EU Common Agricultural Policy, which is to promote agricultural practices that help protect the environment and climate.

The paper [15] outlines a number of positive and negative sustainability considerations associated with the removal of crop residues for expanded uses. As the authors point out, before using crop residues for biofuel production, it should be verified that neutral or positive sustainability impacts can be maintained under site-specific conditions. Crop residues from primary crops are available in significant quantities and do not compete with food production, and to some extent, they are created by virgin cereals production. As a result, there is no need for land conversion. However, this potential largely depends on the development of sustainable and efficient bioenergy systems [15].

Although there are many publications available on the emissions from both biodiesel and rape straw pellets, there is a lack of a comprehensive reference to the impact of the use of these biofuels on GHG emissions in exhaust gases concerning traditional fuels. Therefore, this study aimed to analyze the emissions of selected greenhouse gases (CO_2, CH_4, NO_x) generated during the combustion of different types of biofuels derived from the same plant, i.e., rape (rape oil methyl esters and rape straw pellets), in comparison with conventional fuels (diesel and wood pellets). The obtained results of the emission studies can demonstrate the environmental benefits of using alternative fuels in comparison with their typical counterparts

2. Materials and Methods

2.1. Boiler Tests

The fuels tested were rape straw pellets and A1-grade wood pellets available on the market. The physicochemical properties of rape straw pellets and wood pellets are presented in Table 1.

Table 1. Physicochemical properties of wood pellets and rape straw pellets.

Parameter	Symbol	Unit	Wood Pellets	Rape Straw Pellets
Moisture	W_t^r	%	5.7	9.4
Ash	A^a	%	0.3	10.4
Volatile matter	V^{daf}	%	84.45	64.7
Carbon	C^a	%	49.5	40.1
Hydrogen	H^a	%	6.06	5.8
Sulphur	$S^a{}_A$	%	0.02	0.31
Nitrogen	N^a	%	0.17	0.8
Oxygen *	O	%	38.25	33.19
HHV	$Q_s{}^a$	kJ·kg^{-1}	19,953	15,972
LHV	$Q_i{}^r$	kJ·kg^{-1}	17,893	14,763

* Oxygen was calculated as a complement.

In order to test the combustion of rape straw pellets and wood pellets, a 25 kW boiler adapted for burning pellets was used. The boiler was equipped with a furnace, to which fuel was fed from a reservoir in an automated way. Boiler operation was controlled by a programmed electronic controller. A diagram of the boiler stand is shown in Figure 1.

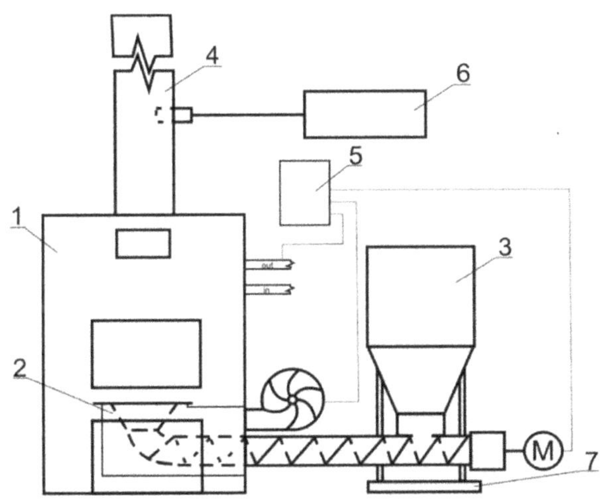

Figure 1. Scheme of boiler stand: 1—test boiler, 2—furnace, 3—pellet reservoir, 4—chimney, 5—boiler controller, 6—exhaust gas analyzer, 7—scales.

Boiler operation was controlled by a programmed electronic controller. The amount of fuel fed for combustion as well as the amount of air required for proper combustion was automatically selected by the controller, based on the results of measurements of the oxygen content in the flue gas provided by the lambda probe and the temperature sensor at the boiler outlet. Combustion tests were carried out under fixed boiler operating conditions at rated settings. Before starting the measurements, the boiler was warmed up for a period of 1 h, the time required to stabilize the boiler was not included in the test duration. The combustion test of individual pellets lasted for 1 h. The fuel consumption was determined by weighing the fuel fed into the reservoir before and after the test for each fuel. The fuel mass flux was, for wood pellets, 6.15 kg·h^{-1} and, for rapeseed straw pellets, 7.63 kg h^{-1}. The flue gas temperature was 138 °C and 134 °C, respectively.

2.2. Engine Testing

B100 biodiesel (fatty acid methyl esters FAME) and ON Efecta Diesel were used for the engine testing (Table 2).

Table 2. Selected physicochemical properties of diesel fuel and methyl esters rape oil.

Parameter	Symbol	Unit	B100	DF
Ester Content	FAME	% (m/m)	98.8	6.8
Density at 15 °C	ρ	kg/m^3	883	835
Viscosity at 40 °C	η	mm^2/s	4.47	2.6
Cetane Number	CN	-	52.1	51.4
Flash point	FP	°C	120	69
Carbon	H^a	% (m/m)	76.9	85.7
Hydrogen	$S^a{}_A$	% (m/m)	11.9	10.6
Oxygen	O	% (m/m)	10.3	2.4
HHV	$Q_s{}^a$	kJ/kg	40,365	45,839
LHV	$Q_i{}^r$	kJ/kg	37,918	43,511

The tests were carried out on a type S-4003 internal combustion engine installed on a dynamometer (Figure 2). The characteristic technical data of the tested engine are presented in Table 3. Despite the successive replacement of the machine park in Polish agriculture with modern tractors characterized by advanced operation and emission parameters, engines like the one under study are still being used to a considerable extent and pose potential ecological problems [16].

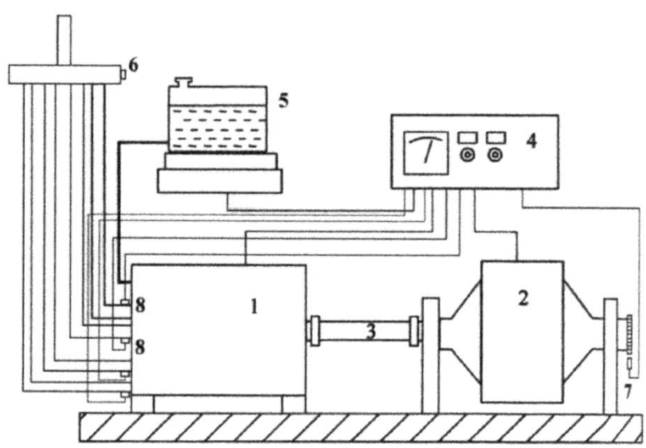

Figure 2. Scheme of the dynamometer stand: 1—test engine, 2—load brake, 3—shaft connecting the engine to the brake, 4—control and measurement system, 5—fuel consumption measuring system, 6—exhaust gas intake, 7—induction speed sensor, 8—exhaust gas temperature sensors.

Table 3. Basic technical data of the S-4003 engine.

Parameter	Unit	Characteristics
Type	-	Self-ignition engine
Cylinder arrangement	-	Vertical in-line
Number of cylinders	-	4
Operating system	-	Four-stroke
Injection system	-	Direct injection
Compression ratio	-	17:1
Engine displacement	dm^3	3.12
Rated power	kW	38.3
Rated speed	rpm	2200
Maximum torque	Nm	186
Maximum torque speed	rpm	1500–1600

Engine load shifting was accomplished with an electric brake type K1-136B-E (asynchronous ring generator), which was also used to start the engine. Emissions were measured based on engine load characteristics at two characteristic speeds (maximum torque and rated power) over the full load range [17,18].

2.3. Measuring Apparatus for Emission Tests

The concentrations of nitrogen oxides (NO_x), carbon dioxide (CO_2), and methane (CH_4) were measured using Testo 350. Testo 350 is a portable exhaust gas analysis system for the measurement of exhaust gas emissions.

2.4. Statistical Analysis

The obtained results were analyzed with the use of analysis of variance (ANOVA). The data analysis was carried out using the Statistica ver. 13 software (TIBCO Software Inc., Palo Alto, CA, USA, 2017) at a significance level of $\alpha < 0.05$.

3. Results and Discussion

3.1. Boiler Test Results

Figures 3–5 show the time courses of changes in CO_2, NO_x, and CH_4 concentrations accompanying the combustion of rape straw pellets and wood pellets. Table 4 lists the average concentrations of the measured flue gas components, calculated from the time courses of their changes. Moreover, Table 5 compares the emissions of these compounds during the combustion of the analyzed fuels in relation to the obtained heat energy.

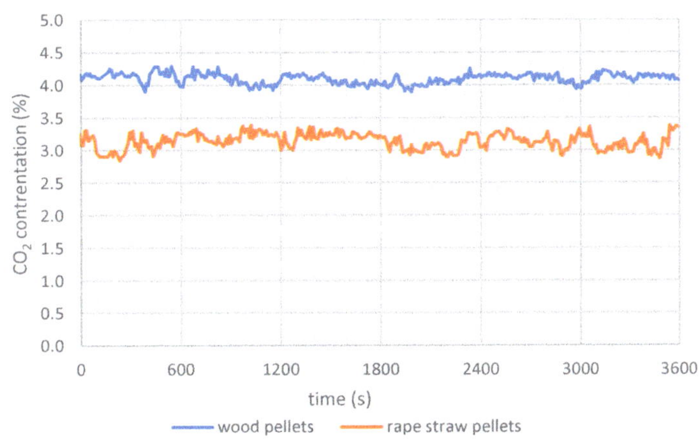

Figure 3. Time course of changes in CO_2 concentration for combustion tests of wood pellets and rape straw pellets.

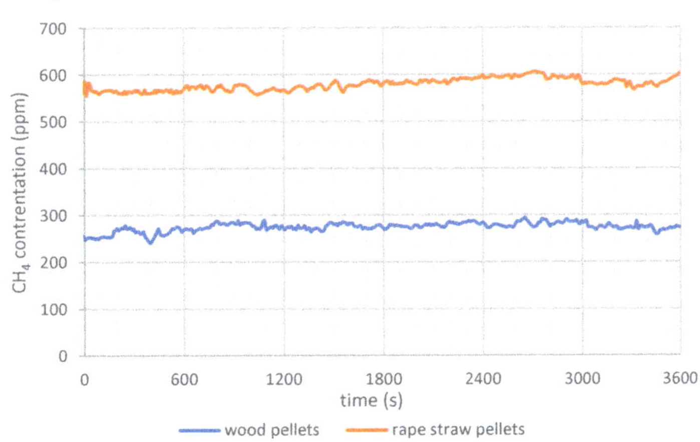

Figure 4. Time course of changes in CH_4 concentration for combustion tests of wood pellets and rape straw pellets.

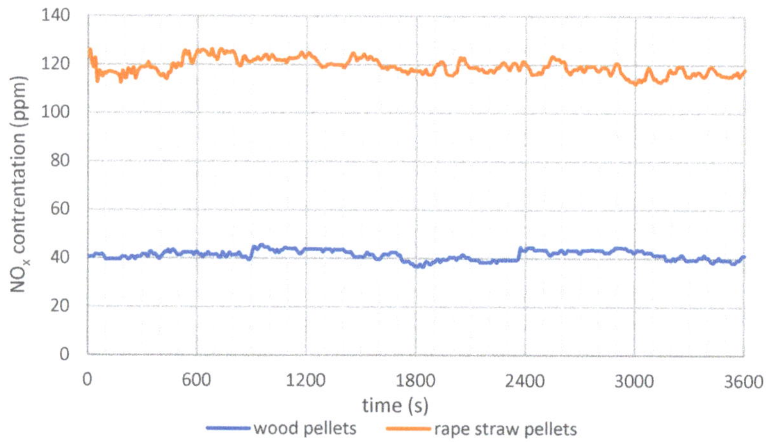

Figure 5. Time course of changes in NO_x concentration for combustion tests of wood pellets and rape straw pellets.

Table 4. Average concentrations of the measured flue gas components.

Specification	Unit	Wood Pellets	Rape Straw Pellets
CO_2	%	4.10	3.15
NO_x	ppm	41.4	119.2
CH_4	ppm	275.3	579.6

Table 5. Average emissions of GHG components ($g{\cdot}kWh^{-1}$).

Emitted Fumes' Component	Units	Wood Pellets	Rape Straw Pellets
CO_2		432	439
NO_x	$g{\cdot}kWh^{-1}$	0.33	0.96
CH_4		1.09	2.29

By comparing the concentrations of the tested gases presented in Figures 3–5 and the average concentrations presented in Table 4, it was found that, in the case of burning pellets from rape straw, the concentrations of CO_2 were lower those of than wood pellets, while the concentrations of NO_x and CH_4 were higher. Especially high concentrations were found for NO_x (almost three times higher) and CH_4 (two times higher for rape straw pellet combustion). The lower values of CO_2 content may be due to the lower calorific value and the higher moisture content of rape straw pellets compared to wood pellets. On the other hand, the higher NO_x concentration was caused by high nitrogen and oxygen contents in rape fuel. The high concentration of CH_4 in the flue gas also indicates an imperfect combustion process—incomplete combustion.

The results of the study presented in Table 5 confirm that the combustion of rape straw pellets is associated with higher NO_x and CH_4 emissions and lower CO_2 emissions in relation to the thermal energy obtained. By comparing the nitrogen content of rape straw pellets (Table 1) with the values typical for wood biomass, which, according to the literature, are below 0.2% [19], it is possible to notice a relatively high content of this element in rape straw pellets. This may be associated with the use of nitrogen-containing mineral fertilizers during rape cultivation, which adversely affects the NO_x emission during the combustion of the investigated biomass. This fact is confirmed by the results of energy and emission studies (NO 421.7 $mg{\cdot}m_n^{-3}$, NO_2 664.8 $mg{\cdot}m_n^{-3}$) (Tables 4 and 5, Figures 2–4).

The NO_x concentration found in the study [20] did not exceed 400 $mg{\cdot}m^{-3}$. If the temperature in the furnace is relatively low, the NO_x concentration depends mainly on the nitrogen stream supplied to the furnace with the fuel, and nitrogen oxides are formed from nitrogen contained in the fuel [21]. In small power boilers, the combustion temperature very often does not exceed 1300 °C, and NO_x are not formed due to the oxidation of atmospheric nitrogen [22–24]. In this case, NO_x emissions should be directly related to the nitrogen content of the fuel.

However, due to the very high ash content in agricultural biomass, when considering the formation of NO_x during its combustion, the catalytic effects of the ash surface must also be taken into consideration. The different NO_x emissions during the combustion of pellets from agricultural biomass may be due to the varying nitrogen content of the biofuel as well as the ash catalyzing the formation of NO_x [25].

Table 6 shows the ANOVA results obtained for the measured emission levels of greenhouse gases, and in Table 7, they were converted to the unit of mass referred to as kWh in boiler tests.

Table 6. ANOVA results for the emission levels (volumetric shares) of GHG components by fuel (rapeseed straw pellets, wood pellets).

GHG Component	Factor	Degrees of Freedom df	Totals of Squares SS	Medium Square MS	Test Function Value F	Calculated Significance Level p
CO_2	fuel	1	163.2181	163.2181	15,593.84	0
NO_x	fuel	1	1,080,143	1,080,143	156,755.5	0
CH_4	fuel	1	16,528,891	16,528,891	151,205.4	0

Table 7. ANOVA results for the emission levels of GHG components ($g \cdot kWh^{-1}$) by fuel (rapeseed straw pellets, wood pellets).

GHG Component	Factor	Degrees of Freedom df	Totals of Squares SS	Medium Square MS	Test Function Value F	Calculated Significance Level p
CO_2	fuel	1	1.1×10^{10}	1.1×10^{10}	2554.695	0
NO_x	fuel	1	70,670,641	70,670,641	37,051.95	0
CH_4	fuel	1	16.279	16.279	24,133.99	0

In both cases, the results of the statistical analysis calculated using the analysis of variance method for all analyzed GHGs due to the type of pellet showed significant differences between the average values (at the significance level of $\alpha = 0.05$). Similar dynamics of changes in the content of the studied flue gas components were observed.

3.2. Engine Test Results

During the tests of the S-4003 engine on a dynamometer bench using FAME fuel and diesel fuel, emissions were measured under various speed-load conditions. Figures 6 and 7 show the waveforms of changes in the level of CO_2 emissions in the exhaust gas of the engine operating according to load characteristics, performed at 1600 rpm and 2200 rpm, respectively.

The study showed a slight decrease in CO_2 emission levels for biofuel, compared to ON, at both engine speeds (Figures 6 and 7). The largest difference between the average values of the concentration of this component in the exhaust gas over the entire engine load range (2.92%), was recorded between B100 and ON at the rated speed (2200 rpm). Figures 8 and 9 show the waveforms of changes in NO_x content in the exhaust gas as a function of engine load (effective power) for 1600 rpm and 2200 rpm, respectively. The study showed that feeding the engine with biodiesel increased in the concentration of nitrogen oxides in the exhaust gas compared to diesel fuel. Averaged over the entire engine load range, the increase in exhaust NO_x concentration was 19.8% at 1600 rpm and 18.9% at 2200 rpm. The level of CH_4 emissions in the exhaust gas (Figures 10 and 11) was found to be significantly lower for the ester-fueled engine compared to the diesel powertrain, averaging over the entire power range by 26.8% at 1600 rpm and by 29.3% at 2200 rpm.

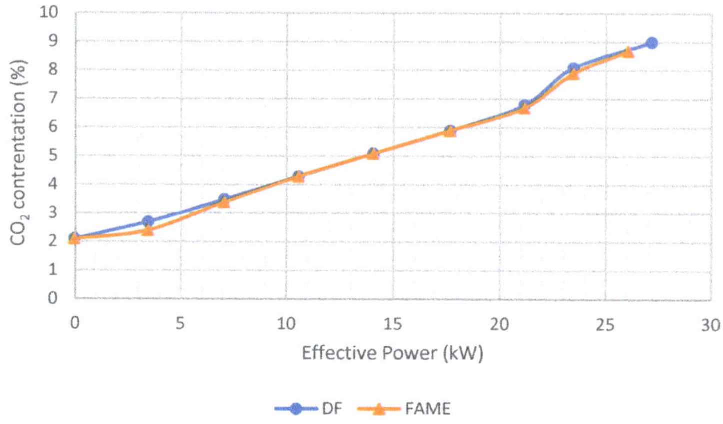

Figure 6. CO_2 emission level in the exhaust gas as a function of the effective power of the S-4003 engine, at an engine speed of 1600 rpm.

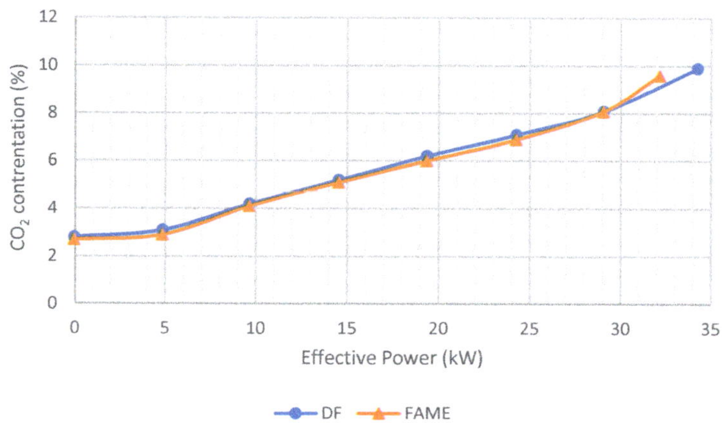

Figure 7. CO_2 emission level in the exhaust gas as a function of the effective power of the S-4003 engine, at an engine speed of 2200 rpm.

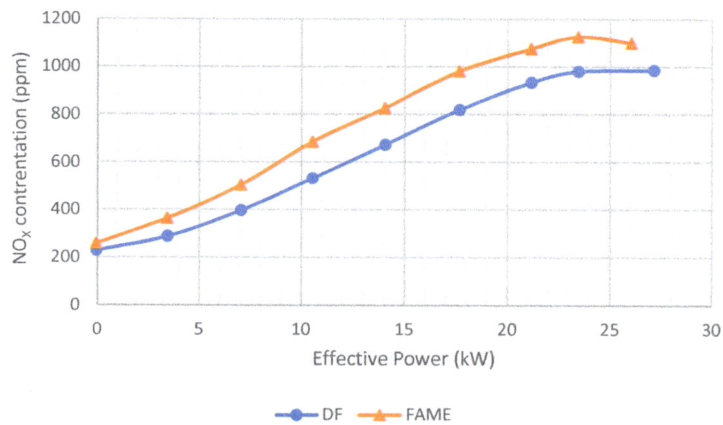

Figure 8. NO_x emission level in the exhaust gas as a function of the effective power of the S-4003 engine, at an engine speed of 1600 rpm.

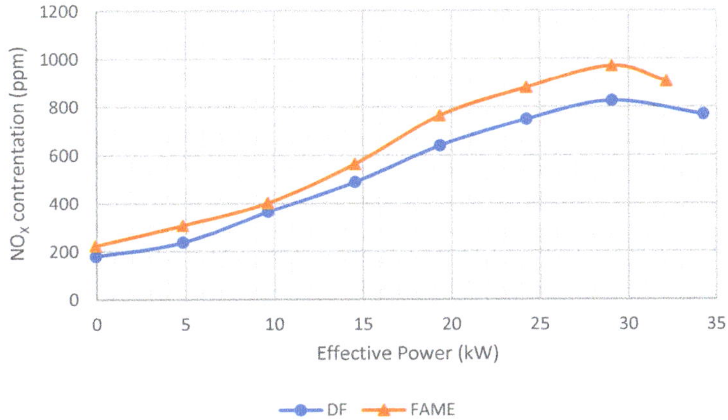

Figure 9. NO_x emission level in the exhaust as a function of the effective power of the S-4003 engine, at an engine speed of 2200 rpm.

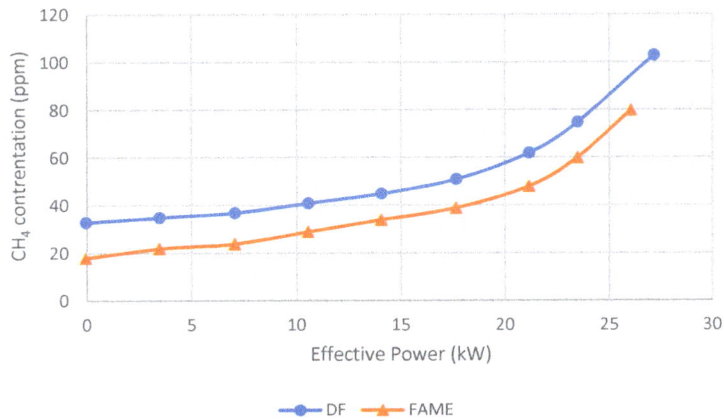

Figure 10. CH_4 emission level in the exhaust as a function of the effective power of the S-4003 engine, at an engine speed of 1600 rpm.

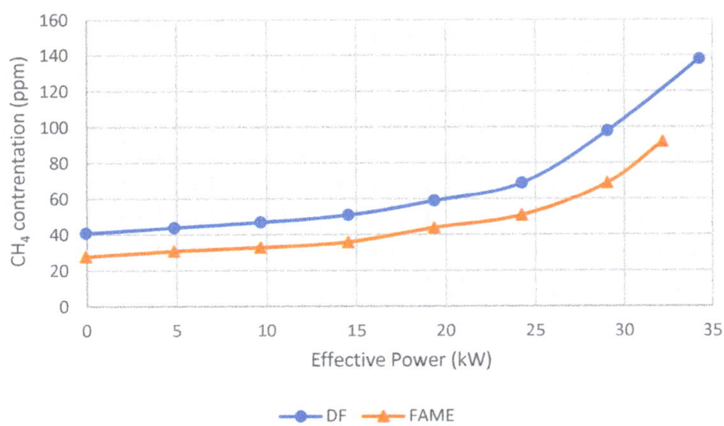

Figure 11. CH_4 emission level in the exhaust as a function of the effective power of the S-4003 engine, at an engine speed of 2200 rpm.

Effective power is one of the most important indicators of an engine's operation, as it determines the amount of energy given up to the consumer at each moment. As the effective power changes, the components of the exhaust gas change significantly, as demonstrated in the study (Figures 6–11). The measurements further showed that feeding the engine with biodiesel, compared to the ON drive, resulted in a decrease in effective power (evident at a full engine load) by 4.1% at 1600 rpm and by 6.1% at 2200 rpm. A particularly important engine operating parameter is effective (overall) efficiency, which characterizes the amount of heat energy supplied to the engine that is converted into useful (effective) work. Internal combustion engines typically have efficiencies of 30–40%. The highest effective efficiency (33%) was achieved by the tested biodiesel-fueled engine in the range of maximum torque characteristics.

GHG emissions are closely related to the temperature inside the cylinders and then, in the exhaust system, to the temperature of the exhaust gas. Exhaust gas temperatures measured during the tests varied from more than 100 °C at no-load engine operation to nearly 600 °C at full-load operation. A slightly higher average exhaust gas temperature, on the order of 1–2%, was recorded when the engine was fueled with biodiesel. Table 8 shows the emissions of measured GHG components converted to g·kWh^{-1} according to the calculation methodology in the works of [26,27].

Table 8. Level of emissions of GHG components (g·kWh^{-1}) due to fuel and engine speed (rpm) and power output (kW).

		B100					ON		
Speed rpm	**Power Output kW**	**CO_2 g·kWh^{-1}**	**NO_x g·kWh^{-1}**	**CH_4 g·kWh^{-1}**	**Speed rpm**	**Power Output kW**	**CO_2 g·kWh^{-1}**	**NO_x g·kWh^{-1}**	**CH_4 g·kWh^{-1}**
1600	0	-	-	-	1600	0	-	-	-
	3.5	2161.0	24.5	3.4		3.5	2063.0	20.9	5.9
	7.1	1275.0	15.4	1.7		7.1	1244.5	12.3	2.6
	10.6	1024.7	13.7	1.4		10.6	998.6	10.4	1.9
	14.1	912.5	12.4	1.2		14.1	896.2	10.0	1.5
	17.7	863.7	12.1	1.1		17.7	881.3	10.3	1.5
	21.2	871.0	11.8	1.2		21.2	868.0	10.0	1.5
	23.5	883.7	10.6	1.3		23.5	881.2	9.0	1.6
	26.1	1019.2	10.8	1.8		27.2	950.1	8.8	2.1
2200	0	-	-	-	2200	0	-	-	-
	4.9	2364.8	21.2	4.9		4.9	2193.8	14.2	6.1
	9.7	1480.0	12.2	2.3		9.7	1400.1	10.3	3.1
	14.6	1195.7	11.1	1.6		14.6	1146.2	9.1	2.2
	19.4	1059.2	11.4	1.5		19.4	1011.5	8.8	1.9
	24.3	988.6	10.6	1.4		24.3	985.2	8.8	1.9
	29.1	1019.5	10.3	1.7		29.1	1001.5	8.6	2.4
	32.2	1106.8	8.8	2.1		34.3	1070.1	7.0	2.9

It can be seen from Table 8 that NO_x and CH_4 exhaust emissions expressed in g·kWh^{-1} showed the same trend of change as the measured values (in ppm), i.e., an increase and a decrease, respectively, in these two exhaust compounds when the engine was fed with B100 biofuel compared to ON. The average over the entire engine power range increase in NO_x was 21.8% at 1600 rpm and 26.5% at 2200 rpm, while the average decrease in CH_4 was 25.4% (1600 rpm) and 25.1% (2200 rpm). In the case of CO_2, however, higher mass emissions were obtained for biodiesel than for diesel, averaging 2.2% at 1600 rpm and 4.0% at 2200 rpm. This is mainly due to the higher fuel consumption and lower heating value of the B100 bioester compared to ON.

Table 9 shows the ANOVA results obtained for the measured GHG emissions in the engine tests.

Table 9. ANOVA results for emission levels (volume shares) of GHG components due to fuel and engine speed (biodiesel, ON).

GHG Component	Factor	Degrees of Freedom df	Totals of Squares SS	Medium Square MS	Test Function Value F	Calculated Significance Level p
CO_2	fuel	1	0.057647	0.057647	0.010419	0.919335
	speed rpm	1	2.359477	2.359477	0.432062	0.515679
NO_x	fuel	1	100,118.4	100,118.4	1.151727	0.291214
	speed rpm	1	140,155.9	140,155.9	1.63585	0.210093
CH_4	fuel	1	2647.059	2647.059	4.195462	0.048811
	speed rpm	1	1152.941	1152.941	1.701444	0.201403

The results of the statistical analysis calculated by the analysis of variance showed a significant effect of fuel type ($p < 0.05$) only for CH_4. For all the other cases analyzed, the differences in the averages are not statistically significant. Table 10 shows the ANOVA results obtained for the emission levels of the studied GHG components converted to $g \cdot kWh^{-1}$ in the engine tests.

Table 10. ANOVA results for the GHG component emissions ($g \cdot kWh^{-1}$) due to fuel and engine speed (biodiesel, ON).

GHG Component	Factor	Degrees of Freedom df	Totals of Squares SS	Medium Square MS	Test Function Value F	Calculated Significance Level p
CO_2	fuel	1	13,402.76	13,402.76	0.070184	0.79301
	speed rpm	1	200,098.5	200,098.5	1.121286	0.298382
NO_x	fuel	1	49.152	49.152	3.352122	0.077781
	speed rpm	1	29.16259	29.16259	1.924114	0.175973
CH_4	fuel	1	3.675	3.675	2.322869	0.138701
	speed rpm	1	2.417044	2.417044	1.537817	0.224884

The results of the statistical analysis calculated using the analysis of variance method for all the cases analyzed (Table 9) showed that the differences in the averages are not statistically significant (at the significance level of $\alpha = 0.05$). A valuable advantage of using biodiesel compared to diesel fuel is the reduced emission of particulate matter and the gaseous components of the exhaust gas (CO, HC), along with the excessive NO_x emissions, as demonstrated in numerous studies, both domestic and foreign. For example, in the work of the authors [28,29], the increase in NO_x concentration for B100 fuel compared to ON was about 10%, confirming the upward trend of this component of the exhaust gas in the completed studies (an increase of about 20%). In turn, the same researchers found a significant decrease in hydrocarbon emissions (about 60%) for B100 fuel compared to ON, which also confirms the results of the present study concerning methane (a decrease in the range of 25–30%). It should be noticed that the performance of an engine depends on its design features (shape of the combustion chamber, design of the fuel injection system, design of the intake system) and operational features (type and characteristics of the fuel, technical condition of individual engine systems, adopted control settings) [30–33]. According to the authors [34], comparing the Life Cycle Assessment (LCA) of biodiesel to that of diesel, the use of this biofuel is more beneficial in terms of reducing the overall greenhouse effect, CO_2 emissions, or carcinogenic compounds. In terms of CO_2, the researchers found that burning each ton of diesel fuel emits 2.8 tons of CO_2 into the atmosphere, while burning biodiesel emits 2.4 tons of CO_2/ton of bioesters. Energy crops are expected to expand significantly in the very short term, bringing significant social and environmental benefits. However, many studies indicate either very positive or negative environmental effects of energy crop cultivation and processing, so there is still a lot of uncertainty regarding these issues [35]. When considering the highest degree of greenhouse gas emission reductions accompanying the use of biomass for energy [36], first-generation biodiesel was found to have less of an impact than first-generation bioethanol concerning bioenergy systems. In addition, for first-generation biodiesel, sunflower showed a lower energy impact than rape. To minimize greenhouse gas emissions from energy systems, an analysis was conducted [37], which indicates that, with adequate biomass availability, liquid fuel production should be based on agricultural residues. Electricity production should be based on forest residues and other woody biomass, and heat production should be based on forest and agricultural residues.

The targets set by the 2009 Renewable Energy Directive for renewable energy make the EU a major global source of demand for biomass. Demand for biomass energy is likely to increase as EU member states set increasingly ambitious renewable energy targets. While biomass power generation is steadily being displaced by other renewable energy sources (mainly wind and photovoltaic power), biomass is likely to remain a major source of renewable heat and transportation biofuels in the short term [38]. The closed-loop economy allows for treating waste biomass as a potential source of valuable energy raw materials. The transition to a closed-loop economy requires, among other things, new ways of transforming hitherto unused waste into new products that constitute resources such as energy. Nevertheless, sustainability criteria should be considered to distinguish raw materials with different climate impacts, as burning different types of biofuels can generate GHG emissions. It is therefore important to control the types of biomass used in order to reduce their negative impact on the climate. Financial and regulatory support should be limited to those raw materials that reduce GHG emissions in the short term, such as lumber residues, agricultural production waste, and post-consumer waste. The overarching goal is to develop sustainable energy systems that do not contribute to further climate change or negatively impact other aspects of sustainability.

4. Conclusions

The novelty of the paper is a comparative study of GHG emissions from the combustion of biodiesel and diesel in an internal combustion engine and from the combustion of various solid biofuels (rapeseed pellets and wood pellets) in a low-power boiler. Such a comparison has not yet been encountered in the literature. As the paper proves, biomass can be used for energy purposes in a variety of ways, and the benefits vary greatly depending on the system used. Bioenergy systems can contribute to climate change mitigation, but the use of biomass resources requires careful consideration of how to target the actions taken in relation to available resources.

The engine test results showed significant reductions in CH_4 emissions when burning B100 biodiesel compared to burning conventional fuel, both at maximum torque and rated speed. Higher NO_x emissions were found for the biofuel burned relative to diesel combustion. In addition, higher CO_2 emissions expressed in $g{\cdot}kWh^{-1}$ were recorded for the combustion of bioester compared to the combustion of diesel, which is mainly due to an increase (about 10%) in the B100 fuel consumption, measured on the dynamometer bench. The engine test results showed a significant reduction in CH_4 emissions when burning B100 biodiesel compared to burning conventional fuel, both at maximum torque and rated speed. The boiler test results indicate that the combustion of rapeseed pellets is associated with higher CH_4 and NO_x emissions compared to the combustion of wood pellets. In contrast, comparable values were found for CO_2 emissions expressed in $g{\cdot}kWh^{-1}$ during the combustion of rapeseed and wood pellets.

Thus, the results obtained from the CO_2 emissivity studies do not clearly indicate the environmental benefits of using the two alternative fuels tested compared to their conventional counterparts. Neither in engine tests nor in boiler tests for NO_x emissions have any ecological benefits been shown from the use of alternative fuels. Only in the case of engine tests were significantly lower CH_4 emissions found when fueling the engine with methyl esters of rapeseed oil.

In conclusion, although the studies did not show significant environmental benefits of using rapeseed-derived fuels, they should not be disqualified. Further research into the combustion process of these fuels can help improve emission factors. In addition, the management of waste biomass such as rapeseed straw, thanks to the possibly longest retention of its economic value, will allow for the sustainable use of rapeseed crops and attempt to close the CO_2 cycle.

The agricultural sector is a sizable emitter of GHGs and a consumer of energy derived mainly from fossil sources. Hence, it is particularly important to use energy (in various forms) from renewable sources in agriculture as much as possible. The results of the authors' research can provide recommendations for the use of, for example, biomass-derived pellets or the more environmentally friendly biodiesel, given the significant amounts of fuel consumed by tractors and other agricultural vehicles. The authors intend to continue this type of research using various forms of agricultural biomass (e.g., biodiesel derived from frying oils) in terms of energy parameters, emissivity, combustion residues, etc.

The research results obtained in this paper are promising and indicate that biomass can play a key role in the diversification of raw material resources and sustainable management based on biotechnology. It is reasonable and interesting to conduct further research on the conversion to energy of different types of waste biomass in terms of greenhouse gas emissions, considering different energy systems. This will allow for an assessment of the environmental impact, selecting and popularizing the best solution.

Author Contributions: Conceptualization, J.W., G.Z., and J.S.-B.; methodology, J.W. and G.Z.; software, G.Z.; validation, A.K.; formal analysis, J.W.; investigation, J.W., G.Z., and J.S.-B.; resources, J.W. and G.Z.; data curation, G.Z.; writing—original draft preparation, J.W., G.Z., and J.S.-B.; writing—review and editing, A.K.; visualization, G.Z.; supervision, A.K.; project administration, G.Z.; funding acquisition, G.Z. All authors have read and agreed to the published version of the manuscript.

Funding: This research received no external funding.

Data Availability Statement: The data presented in this study are available on request from the corresponding author.

Conflicts of Interest: The authors declare no conflict of interest.

References

1. Beach, R.H.; Creason, J.; Ohrel, S.B.; Ragnauth, S.; Ogle, S.; Li, C.; Ingraham, P.; Salas, W. Global Mitigation Potential and Costs of Reducing Agricultural Non-CO_2 Greenhouse Gas Emissions through 2030. *J. Integr. Environ. Sci.* **2015**, *12*, 87–105. [CrossRef]
2. Hanaki, K.; Portugal-Pereira, J. The Effect of Biofuel Production on Greenhouse Gas Emission Reductions. In *Biofuels and sustainability*; Springer: Tokyo, Japan, 2018; pp. 53–71.
3. Oruc, O.; Dincer, I. Environmental Impact Assessment of Using Various Fuels in a Thermal Power Plant. *Int. J. Glob. Warm.* **2019**, *18*, 191–205. [CrossRef]
4. Cho, H.H.; Strezov, V. A Comparative Review on the Environmental Impacts of Combustion-Based Electricity Generation Technologies. *Energy Fuels* **2020**, *34*, 10486–10502. [CrossRef]
5. Jandacka, J.; Caban, J.; Nieoczym, A.; Holubcik, M.; Vrabel, J. Possibilities of Using Wood Waste for the Production of Fuel Briquettes. *Przem. Chem.* **2021**, *100*, 367–374. [CrossRef]
6. Hammar, T. *Climate Impacts of Woody Biomass Use for Heat and Power Production in Sweden*; Swedish University of Agricultural Sciences: Uppsala, Sweden, 2017; ISBN 978-91-576-8872-9.
7. Suhonen, H. *Novel Electrical Particle Emission Reduction Methods for Small-Scale Biomass Combustion*; Finnish Association for Aerosol Research: Helsinki, Finland, 2021; ISBN 978-952-7276-66-2.
8. Houghton, J.T.; Ding, Y.; Griggs, D.J.; Noguer, M.; van der Linden, P.J.; Dai, X.; Maskell, K.; Johnson, C.A. (Eds.) *Climate Change 2001: The Scientific Basis. Contribution of Working Group I to the Third Assessment Report of the Intergovernmental Panel on Climate Change*; Cambridge University Press: Cambridge, UK; New York, NY, USA, 2001; ISBN 0521 01495 6.
9. Estevez, R.; Aguado-Deblas, L.; López-Tenllado, F.J.; Luna, C.; Calero, J.; Romero, A.A.; Bautista, F.M.; Luna, D. Biodiesel Is Dead: Long Life to Advanced Biofuels—A Comprehensive Critical Review. *Energies* **2022**, *15*, 3173. [CrossRef]
10. Dzieniszewski, G.; Kuboń, M.; Pristavka, M.; Findura, P. Operating Parameters and Environmental Indicators of Diesel Engines Fed with Crop-Based Fuels. *Agric. Eng.* **2021**, *25*, 13–28. [CrossRef]
11. Lee, D.; Pomraning, E.; Rutland, C.J. LES Modeling of Diesel Engines. *SAE Trans.* **2002**, *11*, 2566–2578. [CrossRef]
12. Król, D.; Poskrobko, S.; Tokarz, Z.; Gościk, J.; Wasiak, A. The Fuel Biomass about Raised Caloricity. *Arch. Waste Manag. Environ. Prot.* **2017**, *19*, 11–16.
13. Juszczak, M. Evaluation of CO, NO, NO_x and Dust Concentration Values in Flue Gas from Thermal Conversion of Straw Ballots. *Arch. Gospod. Odpad. Ochr. Śr.* **2010**, *12*, 1–14.
14. Kachel, M.; Kraszkiewicz, A.; Subr, A.; Parafiniuk, S.; Przywara, A.; Koszel, M.; Zając, G. Impact of the Type of Fertilization and the Addition of Glycerol on the Quality of Spring Rape Straw Pellets. *Energies* **2020**, *13*, 819. [CrossRef]

15. Battaglia, M.; Thomason, W.; Fike, J.H.; Evanylo, G.K.; von Cossel, M.; Babur, E.; Iqbal, Y.; Diatta, A.A. The Broad Impacts of Corn Stover and Wheat Straw Removal for Biofuel Production on Crop Productivity, Soil Health and Greenhouse Gas Emissions: A Review. *GCB Bioenergy* **2021**, *13*, 45–57. [CrossRef]
16. Lorencowicz, E.; Uziak, J. Regional Structure of Tractor Market in Poland. *Agric. Eng.* **2020**, *24*, 51–62. [CrossRef]
17. Wasilewski, J.; Szyszlak-Bargłowicz, J.; Zając, G.; Szczepanik, M. Assessment of CO_2 Emission by Tractor Engine at Varied Control Settings of Fuel Unit. *Agric. Eng.* **2020**, *24*, 105–115. [CrossRef]
18. Wasilewski, J.; Krasowski, E. *Internal Combustion Engines*; Wydawnictwo Uniwersytetu Przyrodniczego: Lublin, Poland, 2015; ISBN 83-7259-238-1.
19. Demirbas, A. Combustion Characteristics of Different Biomass Fuels. *Prog. Energy Combust. Sci.* **2004**, *30*, 219–230. [CrossRef]
20. Król, D.; Łach, J.; Poskrobko, S. O Niektórych Problemach Związanych z Wykorzystaniem Biomasy Nieleśnej w Energetyce. *Energetyka* **2010**, *1*, 53–62.
21. Juszczak, M. Concentrations of Carbon Monoxide and Nitrogen Oxides from a 15 KW Heating Boiler Supplied Periodically with a Mixture of Sunflower Husk and Wood Pellets. *Environ. Prot. Eng.* **2014**, *40*, 66–74. [CrossRef]
22. Zhao, W.; Li, Z.; Wang, D.; Zhu, Q.; Sun, R.; Meng, B.; Zhao, G. Combustion Characteristics of Different Parts of Corn Straw and NO Formation in a Fixed Bed. *Bioresour. Technol.* **2008**, *99*, 2956–2963. [CrossRef] [PubMed]
23. Houshfar, E.; Skreiberg, Ø.; Løvås, T.; Todorović, D.; Sørum, L. Effect of Excess Air Ratio and Temperature on NO_x Emission from Grate Combustion of Biomass in the Staged Air Combustion Scenario. *Energy Fuels* **2011**, *25*, 4643–4654. [CrossRef]
24. Li, Z. *Corn Straw and Biomass Blends: Combustion Characteristics and NO Formation*; Nova Science Publishers: New York, NY, USA, 2009; ISBN 1-61122-445-4.
25. Verma, V.K.; Bram, S.; Delattin, F.; Laha, P.; Vandendael, I.; Hubin, A.; De Ruyck, J. Agro-Pellets for Domestic Heating Boilers: Standard Laboratory and Real Life Performance. *Appl. Energy* **2012**, *90*, 17–23. [CrossRef]
26. Sarkan, B.; Kuranc, A.; Sejkorova, M.; Caban, J.; Loman, M. Comparison of the Exhaust Emissions of Heavy-Duty Vehicle Engines Powered by Diesel Fuel (DF) and Natural Gas (LNG) in Real Operation Conditions. *Przem. Chem.* **2022**, *101*, 37–41. [CrossRef]
27. Kuranc, A.; Słowik, T.; Krzaczek, P.; Maj, G. Emission of Fumes of Ursus MF235 under Conditions of Load with the Use of a Movable Dynamometric Stand. *Agric. Eng.* **2016**, *20*, 101–112. [CrossRef]
28. Maia, E.C.R.; Borsato, D.; Moreira, I.; Spacino, K.R.; Rodrigues, P.R.P.; Gallina, A.L. Study of the Biodiesel B100 Oxidative Stability in Mixture with Antioxidants. *Fuel Process. Technol.* **2011**, *92*, 1750–1755. [CrossRef]
29. Demirbas, A. Progress and Recent Trends in Biodiesel Fuels. *Energy Convers. Manag.* **2009**, *50*, 14–34. [CrossRef]
30. Arshad, M.; Zia, M.A.; Shah, F.A.; Ahmad, M. An Overview of Biofuel. In *Perspectives on Water Usage for Biofuels Production: Aquatic Contamination and Climate Change*; Arshad, M., Ed.; Springer International Publishing: Cham, Switzerland, 2018; pp. 1–37. ISBN 978-3-319-66408-8.
31. Kousoulidou, M.; Fontaras, G.; Ntziachristos, L.; Samaras, Z. Biodiesel Blend Effects on Common-Rail Diesel Combustion and Emissions. *Fuel* **2010**, *89*, 3442–3449. [CrossRef]
32. Zajac, G.; Wegrzyn, A. Analysis of Work Parameters Changes of Diesel Engine Powered with Diesel Fuel and FAEE Blends. *Eksploat. Niezawodn.-Maint. Reliab.* **2008**, *38*, 17–24.
33. Silitonga, A.S.; Hassan, M.H.; Ong, H.C.; Kusumo, F. Analysis of the Performance, Emission and Combustion Characteristics of a Turbocharged Diesel Engine Fuelled with Jatropha Curcas Biodiesel-Diesel Blends Using Kernel-Based Extreme Learning Machine. *Environ. Sci. Pollut. Res.* **2017**, *24*, 25383–25405. [CrossRef] [PubMed]
34. Nanaki, E.A.; Koroneos, C.J. Comparative LCA of the Use of Biodiesel, Diesel and Gasoline for Transportation. *J. Clean. Prod.* **2012**, *20*, 14–19. [CrossRef]
35. Fazio, S.; Monti, A. Life Cycle Assessment of Different Bioenergy Production Systems Including Perennial and Annual Crops. *Biomass Bioenergy* **2011**, *35*, 4868–4878. [CrossRef]
36. Thornley, P.; Gilbert, P.; Shackley, S.; Hammond, J. Maximizing the Greenhouse Gas Reductions from Biomass: The Role of Life Cycle Assessment. *Biomass Bioenergy* **2015**, *81*, 35–43. [CrossRef]
37. Bentsen, N.S.; Jack, M.W.; Felby, C.; Thorsen, B.J. Allocation of Biomass Resources for Minimising Energy System Greenhouse Gas Emissions. *Energy* **2014**, *69*, 506–515. [CrossRef]
38. Brack, D.; Hewitt, J.; Marchand, T.M. Woody Biomass for Power and Heat: Demand and Supply in Selected EU Member States; United Kingdom. 2018. Available online: https://policycommons.net/artifacts/613609/woody-biomass-for-power-and-heat/1593617/ (accessed on 15 September 2022).

energies

Article

Modeling and Control Strategy of Wind Energy Conversion System with Grid-Connected Doubly-Fed Induction Generator

Abrar Ahmed Chhipą [1], Prąsun Chakrabarti [2], Vadim Bolshev [3,*], Tulika Chakrabarti [4], Gennady Samarin [3,5], Alexey N. Vasilyev [3], Sandeep Ghosh [6] and Alexander Kudryavtsev [3]

1 Techno India NJR Institute of Technology, Udaipur 313003, Rajasthan, India
2 Department of Computer Science Engineering, ITM (SLS) Baroda University, Vadodara 391510, Gujarat, India
3 Federal Scientific Agroengineering Center VIM, 109428 Moscow, Russia
4 Department of Chemistry, Sir Padampat Singhania University, Udaipur 313601, Rajasthan, India
5 Department of Agricultural Power Engineering, Northern Trans-Ural State Agricultural University, 625003 Tyumen, Russia
6 Department of Electrical Engineering, Indian Institute of Technology (BHU), Varanasi 221005, Uttar Pradesh, India
* Correspondence: vadimbolshev@gmail.com; Tel.: +7-499-174-85-95

Abstract: The most prominent and rapidly increasing source of electrical power generation, wind energy conversion systems (WECS), can significantly improve the situation with regard to remote communities' power supply. The main constituting elements of a WECS are a wind turbine, a mechanical transmission system, a doubly-fed induction generator (DFIG), a rotor side converter (RSC), a common DC-link capacitor, and a grid-side converter. Vector control is center for RSC and GSC control techniques. Because of direct and quadrature components, the active and reactive power can also be controller precisely. This study tracks the maximum power point (MPP) using a maximum power point tracking (MPPT) controller strategy. The MPPT technique provides a voltage reference to control the maximum power conversion at the turbine end. The performance and efficiency of the suggested control strategy are validated by WECS simulation under fluctuating wind speed. The MATLAB/Simulink environment using simpower system toolbox is used to simulate the proposed control strategy. The results reveal the effectiveness of the proposed control strategy under fluctuating wind speed and provides good dynamic performance. The total harmonic distortions are also within the IEEE 519 standard's permissible limits which is also an advantage of the proposed control approach.

Keywords: wind energy conversion system; doubly-fed induction generator; MPPT; vector control; renewable energy; WECS; DFIG

Citation: Chhipą, A.A.; Chakrabarti, P.; Bolshev, V.; Chakrabarti, T.; Samarin, G.; Vasilyev, A.N.; Ghosh, S.; Kudryavtsev, A. Modeling and Control Strategy of Wind Energy Conversion System with Grid-Connected Doubly-Fed Induction Generator. *Energies* **2022**, *15*, 6694. https://doi.org/10.3390/en15186694

Academic Editor: King Jet Tseng

Received: 16 August 2022
Accepted: 9 September 2022
Published: 13 September 2022

1. Introduction

Energy has been a remarkable source and a prominent need for the technological evolution of human race. Power generation systems have grown in size in order to meet the increasing electricity demand of the 21st century. Urbanization and expeditious growth in population has engendered a global energy shortage. On the basis of exhaustibility, power generation systems are categorized under two major groups known as renewable energy systems and non-renewable energy systems. Renewable energy is acquired from natural resources such as the sun, biofuel, wind, water stream, geothermal and tides, whereas non-renewable energy is derived from finite sources like nuclear fuels and fossil fuels (coal, crude oil and natural gas) [1].

With the increase in electrical energy demand and reduction of conventional energy sources, it is necessary to increase electrical energy generation through renewable energy sources like wind, solar, etc. [1,2]. Even today, fossil fuels like coal and gas-based electric power generation dominate worldwide, with steam power plants covering a significant part of per-capita electricity generation. The power systems based on fossil fuels thus form the primary source of energy and seek continuous improvements in different areas such as reliability, stability, controllability, and environmental aspects. Inadequate quality control, aided by high maintenance costs, contributes to the increase in energy bills to the utility and end consumer. Further, fossil fuels such as coal, gas, and petroleum are limited reserves and produce harmful air pollution [3,4]. In addition, nuclear power plants produce harmful radioactive waste, e.g., isotopes of plutonium which cause health hazards to humans and other living bodies including agricultural ones. Also, the maintenance cost of the nuclear power plant is much more than any other power station [5]. Apart from the harmful wastes, environmental issues, high operational costs, and maintenance issues, fossil fuels are limited in quantity and shall vanish one day [6].

Fortunately, nature has provided us alternate energy sources such as wind, solar, geothermal tidal, and fuel cell, which are available abundantly in rural areas and shall be there forever. Wind energy remains a prominent and cleaner energy source among all candidates for naturally available green energy due to its emission-free nature [7].

Wind energy conversion systems (WECS) have become the fastest-growing source of electrical power generation globally and shall cover a significant share of global electricity capacity in the future. Developing countries can quickly adopt this technology owing to its lesser complexity in design, manufacturing, and installation.

The wind turbine converts the kinetic wind power stroked on the rotor blades to mechanical rotational energy. Based on the rotor speed, Wind turbines are categorized as fixed speed wind turbines (FSWT) and variable speed wind turbines (VSWT). The rotational speed is preserved as constant for every wind speed in the FSWT system. The VSWT has the ability to vary their rotor speed to succeed the instantaneous wind speed variations. Generally, the variable-speed turbines are more engrossing than the fixed-speed turbines where the speed of wind varies significantly. There are various reasons for using VSWT over the unreliable operating speed in WECS, these include reduced stress in mechanical parts, reduction in acoustic noise and exalted power quality.

DFIG-based WECS are preferred and are employed worldwide. Such systems consist of a slip ring induction motor or wound rotor induction machine as an electrical power conversion device known as a doubly-fed induction generator [8–10]. They are controlled with electronic converters, making it possible to control the speed of the rotor and power. In the wound rotor induction machine, the stator is directly connected to the grid, and back-to-back connected converters connect rotor windings. It is known as DFIG as it allows the flow of electrical energy in both directions; into the grid when generator operating speed is super-synchronous and into the rotor, if generator speed is sub-synchronous [11]. The bidirectional AC/DC/AC converters, which are connected in between rotor circuit and grid, control the speed above the synchronous speed, and the power is generated from both stator and rotor [12,13].

The main advantages of DFIG are as follow [14,15]:

- Able to supply the power at constant voltage and frequency while the speed of the rotor varies.
- Improve the efficiency of wind generator, rotor speed may vary according to wind speed.
- Power electronic converters require lower power ratings as they have to handle a fraction of the total power.
- Independent control of active and reactive power is possible, and so the power factor can be controlled [16].

The effectiveness of a DFIG-based wind system has been gauged higher than the other wind power generators and so it is an agreeable option for grid-connected wind energy systems driven by VSWT. By considering the favourable features of a DFIG machine, this research work dealt with a DFIG-based VSWT, for wind power applications.

The main power flows out of the stator and is fed to the grid through the generator and transformer. The power handling capacity of the back-to-back converter is about 30%

of the rated power. This back-to-back converter works as a bisectional power flow. The rotor power is first converted to DC by rotor side converter (RSC) [17], then this DC is again converted to AC utilizing grid side converter (GSC) [18] and fed to the grid via the transformer. In wind speed variation, these converters prevent the machine from damage and support the grid line [19]. The RSC is responsible for the active, reactive power exchange to the grid. GSC controls the DC link voltage and grid point of the common coupling power factor utilizing reactive power exchange [20].

Climatic changes and their uncertain nature are major factors that determine the reliability of wind energy. Despite these above-mentioned factors, wind energy system retains some more challenges for researchers, including utility grid integration and the locus of wind turbine. Since the power obtained from the WECS is directly proportional to the wind speed, even a slight variation in wind speed leaves a strong impression in the extracted power. Also, the electrical grid should be extended at constant amplitude and frequency; hence the evoked wind power will be incompatible with the utility grid [21,22].

The unbalanced grid voltage may shrink the span of gearbox and DC-link capacitor owing to plentiful electric torque pulsation. Further, inadequate damped oscillation in the course of voltage sags leads to inferior low-voltage ride through capability. Direct power control, AC grid power fluctuations, direct torque control and decoupled control of the active and reactive power are some additional concerns that are to be considered with regard to WECS [23]. Because of these, some appropriate control schemes have to be employed in order to harvest the maximum power and incur the constant voltage in wind energy applications.

Wind power generation depends on geographical and weather conditions. Thus, developing such advanced types of control strategies can make the system work with maximum efficiency and produce optimum power considering grid codes and IEEE standards [24]. Our study focuses on vector control based on the voltage and flux vector in *dq* reference frame, which provides independent control of active and reactive power.

This paper is organized as follows: Section 1 discusses the importance of the use of DFIG-based WECS. Section 2 presents a literature survey in terms of technology development in the field of wind energy conversion system. In Section 3 we present the system configuration of grid-connected DFIG-based WECS and the modeling of wind turbine and doubly-fed induction generators. Section 4 describes the MPPT controller for optimum torque reference generation in RCS. Section 5 presents the vector control strategy for RCS and GCS, which are based on PI controller. At last, in Sections 6 and 7, we conclude the study by a simulation analysis and simulation results of the proposed system control with variable input wind speed.

2. Literature Survey

This section deals with the literature survey in terms of state of the art and significant developments in the area of wind energy conversion systems. It is based on its classification into two broad categories: fixed speed wind turbine- and variable speed wind turbine-based generating systems. The literature review on the use of a battery energy storage system with wind energy conversion systems is also carried out in detail. In the past, the most commonly used generators were the squirrel cage induction generator (SCIG) in terms of wind turbine generators (WTGs) using fixed speed wind turbines. The use of an isolated asynchronous generator (IAG) for the stand-alone power generation has been proven promising for the last few decades due to its simplicity, brushless construction, ease in maintenance, inherent short circuit protection and large torque/weight ratio [25,26]. Another attempt has been made by Quazene and McPherson [27] to analyze the operation of an IAG in stand-alone WECS. They used an IAG feeding with resistive loads where frequency and voltage were not regulated across the loads. The main barriers in IAG commercial adoption are its poor voltage and frequency regulation characteristics. Further, when the load is directly connected across the IAG bus, the IAG frequency depends on the prime-mover speed which is a function of input wind power and connected consumer loads [28]. Casielles et al. [28] have reported a control system for a wind turbine driven IAG. In these works, the authors used discrete switching for the number of switched capacitors and resistors to balance active and reactive powers as the wind speed changes. However, the discrete switching of capacitors and resistors has provided an inferior performance of the controller as far as power quality, proficiency and maintenance. It has been shown that a voltage source converter (VSC) facilitates the achievement of better system behavior in terms of voltage regulation, reactive power compensation and frequency stabilization. Kasal and Singh [29] have reported different system configurations of voltage and frequency controllers for an IAG and their design for stand-alone WECS. They compared a DFIG-based, grid-fed WECS with other WECS technologies such as fixed speed and a sensor-less vector control techniques for a DFIG-based stand-alone power generation utilizing model reference adaptive system (MRAS) observer. The sensor-less plan has been approved under both steady-state and transient conditions utilizing exploratory outcomes. Pena et al. [30] have discovered a control scheme for the stand-alone operation of a DFIG supplying inequivalence load. The front-end converter was controlled to compensate for inequivalence loads by supplying positive and negative distributed load currents. Jain and Ranganathan [31] discussed a vector-controlled scheme for back-to-back connected VSC's for a DFIG-based stand-alone power generation. In this sensorless control scheme, the function of active filtering was introduced for a DFIG based stand-alone energy conversion system. Harmonics present in the consumer loads were compensated through a front-end converter. Goel et al. [32] announced that two DFIGs connected parallelly worked for stand-alone WECS feeding nearby loads. In this control, the authors utilized three VSC's for the control of two parallel operated DFIG's. The BESS was utilized in the middle of the DC-link of three VSC's for load levelling. Variable speed wind turbines are presently utilized as a part of WECS innovation. The variable speed activity is possible because the power that the electronics converters provide allows for complete decoupling from the grid. Full rating variable-speed WECS are flexible in terms of different types of generators used in these WECSs. Watson et al. [33] proposed an IAG-based variable speed WECS for controllable DC control supply. A three-phase controlled rectifier was utilized at the generator side to change variable voltage/variable frequency AC supply to a constant supply DC voltage. The reactive power demand of an IAG was met by using the self-excitation capacitors and the proper controls variable speed IAG-based WECS technologies was done. It has been proven through the obtained results that a DFIG-based WECS has been suggested due to its enhancement of the wind energy capturing capacity for the same rating of another machine in WECS [34]. The DFIG is the only machine that can give more than its rated power without being overloaded [35,36]. The DFIG has the capability to produce power under the change in prime-mover speed at constant voltage and frequency. Further, the DFIG can be controlled from the rotor side and it reduces the overall controller rating (i.e., the fraction of the DFIG rating) [37]. These features make a DFIG one of the most attractive choices for variable speed prime-movers. Yamamoto and Motoyoshi [36] have reported that the DFIG-based energy conversion system controls the active and reactive power of the system through back-to-back connected VSC's feeding the generated power to the grid. The literature of different control strategies of a DFIG feeding generating power to the grid is also addressed [37–45]. Some literature is also available on the sensorless operation of a DFIG supplying generating power to the grid [46–48]. In [49], the authors reported an implementation of a vector-controlled DFIG based stand-alone WECS. The execution of the voltage and frequency controller has been shown with linear loads. Cardenas et al. [50] examined how over-firing angles and electrical loads at a DC bus have been proposed for constant DC bus voltage. In [51], authors reported a rule-based fuzzy logic controller for a variable speed IAG-based stand-alone wind energy conversion system. Simoes et al. [52] described the control strategy, design and performance evaluation of a fuzzy logic-based variable speed wind energy conversion using an IAG. Poddar et al. [53] reported a variable speed controller for a 225 kW IAG based grid integrated wind energy conversion system using a sensorless control algorithm that was realized using a direct torque control method.

This control algorithm was based on active and reactive power control. In [54] the authors suggest proposing an automatic mode switch control strategy for IAG-based small wind turbines under conditions of grid failure in the stand-alone mode of a grid-connected system. In [55,56] the authors describe a phase-locked loop (PLL) control configuration for a vector-controlled capacitor-excited IAG that uses PWM controlled VSC converters for AC and DC voltage regulation in variable speed WECS. Barakati et al. [57] propose a grid converter for an IAG-based variable speed WECS integrated with the grid. In this work, the authors propose a single-stage conversion in place of two-stage conversion with AC-DC-AC converters. The effectiveness of the controller has been shown through simulation results. Agrawal et al. [58] have reported a novel maximum power tracking algorithm using a grid converter for an IAG-based WECS. Vas and Li [59] have developed a package for simulating the performance of a vector-controlled generator and named it SIMUVEC, applicable for both voltage source and current source inverter fed induction motors. Gabriel et al. [60] have presented a field-oriented control of AC motors using microprocessors. The change of the transient stability margin of the network wi th increment in wind power penetration when the wind farms are associated with the distinguished areas, whereas the stability condition deteriorates with increased penetration when the DFIG is connected to with some other location. Yuan-Kang Wu et al. [61] presented different control strategies in the rotor side converter of a 2 MW DFIG wind turbine. these control strategies are operated on MPPT mode to obtain maximum efficiency. These are categories such as voltage-oriented control, flux oriented control, direct torque control and direct power control. The performance was obtained under these strategies and simulated by PSCAD/EMTDC software[61].

3. System Configuration and Modeling of DFIG

The system configuration of a grid-connected DFIG-based WECS is shown in Figure 1. This configuration is designed to deliver the power of 2 MW. The system comprises a wind turbine, DFIG, rotor-side filter, grid side filter, MPPT controller, DC-link capacitor, and back-to-back three-phase pulse width modulated (PWM) VSCs with their controller. The VSC connected to the rotor winding through the rotor filter is named RSC. The stator winding is directly connected to the grid, and the VSC is connected at the point of coupling (PoC) in GSC [62]. The RSC and GSC are responsible for achieving the different operating conditions of DFIG. The block scheme of the proposed system is shown in Figure 2.

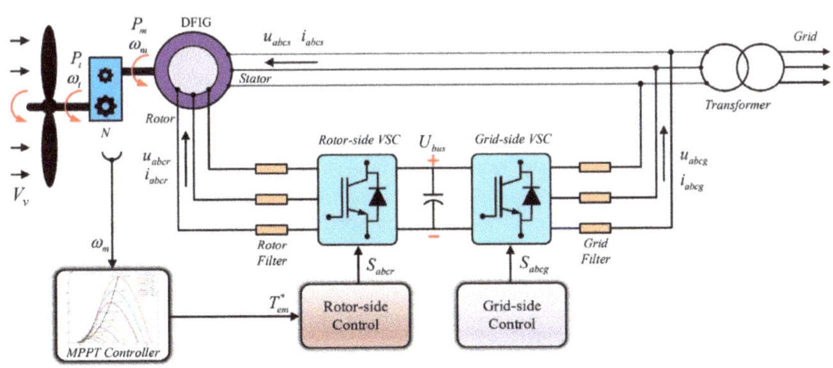

Figure 1. Configuration of variable speed wind turbine with DFIG.

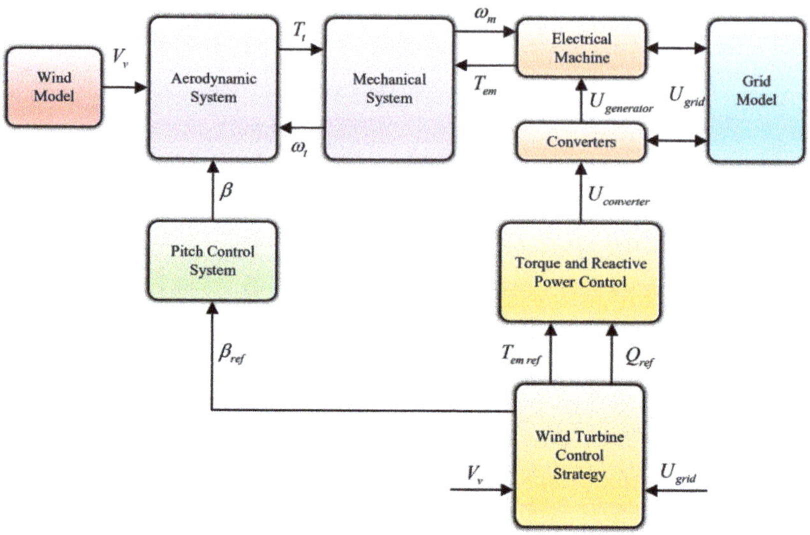

Figure 2. Block diagram of a grid-connected DFIG.

3.1. Wind Turbine Modeling

The kinetic energy of the wind is converted to rotational energy in the form of a torque by the wind turbine (WT). The power available in the wind is given by Equation (1):

$$P_v = \frac{1}{2}\rho A V_v^3 \tag{1}$$

where

A: Area swept by turbine blades (m^2).
ρ: The air density (kg/m^3).
V_v: Wind speed (m/s).

The power extracted by the turbine from the available power in the wind is given by Equation (2):

$$P_t = \frac{1}{2}\rho\pi R^2 V_v^3 C_p(\lambda,\beta) \tag{2}$$

where

R: The radius of turbine rotor (m)
$C_p(\lambda,\beta)$: The power coefficient
C_p is a function of tip speed ratio (λ) and the pitch angle (β).

$C_p(\lambda,\beta)$ is expressed by Equation (3) and Equation (4):

$$C_p(\lambda,\beta) = c_1\left(\frac{c_2}{\lambda_i} - c_3\beta - c_4\beta^{c_5} - c_6\right).e^{\frac{-c_7}{\lambda_i}} \tag{3}$$

where $c_1 = 0.73;\ c_2 = 151;\ c_3 = 0.58;\ c_4 = 0.002;\ c_5 = 2.4;\ c_6 = 13.2;\ c_7 = 18.4$

$$C_p(\lambda,\beta) = 0.73.\left(\frac{151}{\lambda_i} - 0.58\beta - 0.002\beta^{2.4} - 13.2\right).e^{\frac{-18.4}{\lambda_i}} \tag{4}$$

$$\lambda_i = \frac{1}{\lambda + 0.02\beta} - \frac{0.003}{\beta^3 + 1} \tag{5}$$

According to the Betz limit, the maximum theoretical value C_p is expressed in Equation (6):

$$C_{p\ Theo\ \max} = 0.593 = 59.3\% \tag{6}$$

The λ is expressed in Equation (7):

$$\lambda = \frac{R\omega_t}{V_v} \tag{7}$$

The turbine generated torque is expressed in Equation (8):

$$T_t = \frac{P_t}{\omega_t} \tag{8}$$

ω_t: The angular rotational speed of wind turbine rotor (rad/sec).

3.2. DFIG Modeling

The DFIG is composed of stator and rotor windings. It features slip rings. Three-phase insulated windings are mounted on the stator connected to the grid through a three-phase transformer. The rotor is also built of three-phase insulated windings in the same way as the stator. A set of slip rings and brushes connects the rotor windings to an external stationary circuit. These components allow for either injection into or absorption from the rotor windings of the control rotor current [62–65].

The direct and inverse transformation is used to represent the dynamic model of the DFIG. Using space vector theory, the three windings of the rotor and stator can be represented by two winding $\alpha\beta$ as stationary for stator and winding dq as rotating for the rotor.

The stator and rotor voltage vector is expressed as:

$$\vec{u}_s \Rightarrow \begin{cases} u_{ds} = R_s i_{ds} + \frac{d\psi_{ds}}{dt} - \omega_s \psi_{qs} \\ u_{qs} = R_s i_{qs} + \frac{d\psi_{qs}}{dt} + \omega_s \psi_{ds} \end{cases} \tag{9}$$

$$\vec{u}_s \Rightarrow \begin{cases} u_{dr} = R_r i_{dr} + \frac{d\psi_{dr}}{dt} - \omega_r \psi_{qr} \\ u_{qr} = R_r i_{qr} + \frac{d\psi_{qr}}{dt} + \omega_r \psi_{dr} \end{cases} \tag{10}$$

where u_{ds}, u_{qs}, u_{dr}, and u_{qr}: stator and rotor voltages in the dq frame, respectively. i_{ds}, i_{qs}, i_{dr}, and i_{qr}: stator and rotor current in the dq frame, respectively.R_r, R_s, ω_s and ω_r: stator and rotor phase resistances and angular velocity, respectively. From Equations (9) and (10), dq equivalent electric circuit is represented by Figure 3.

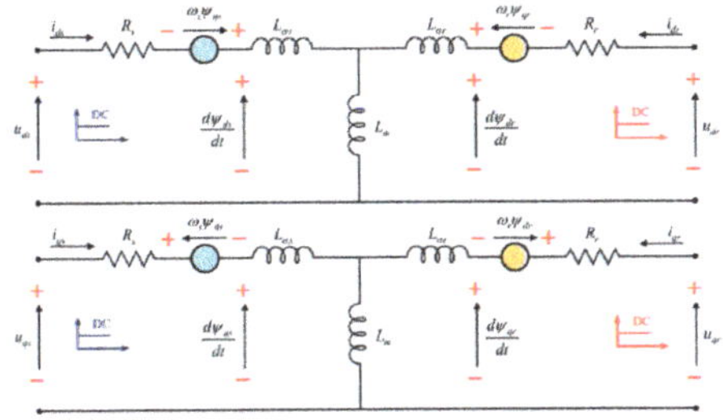

Figure 3. DFIG dq equivalent circuit.

The stator and rotor flux vector are expressed in Equation (11) and Equation (12), respectively:

$$\vec{\psi}_s \Rightarrow \begin{cases} \psi_{ds} = L_s i_{ds} + L_m i_{dr} \\ \psi_{qs} = L_s i_{qs} + L_m i_{qr} \end{cases} \tag{11}$$

$$\vec{\psi}_r \Rightarrow \begin{cases} \psi_{dr} = L_m i_{ds} + L_r i_{dr} \\ \psi_{qr} = L_m i_{qs} + L_r i_{qr} \end{cases} \tag{12}$$

where $\vec{\psi}_s$, $\vec{\psi}_r$ are the flux vectors for stator and rotor, respectively. ψ_{ds}, ψ_{qs} are the fluxes along the dq axis stator. ψ_{dr}, ψ_{qr} are the fluxes along with the dq axis rotor. L_s, L_r: stator and rotor phase leakage inductances, respectively, L_m: stator–rotor mutual inductance, and p: is the number of pole pairs of the generator.

The expression of electromagnetic torque is expressed in Equation (13):

$$T_{em} = \frac{3}{2} p \frac{L_m}{L_s} (\psi_{qs} i_{dr} - \psi_{ds} i_{qr}) \tag{13}$$

The active and reactive power equations of the stator and rotor are given by Equation (14) and Equation (15):

$$\begin{cases} P_S = \frac{3}{2}(u_{ds} i_{ds} + u_{qs} i_{qs}) \\ Q_S = \frac{3}{2}(u_{qs} i_{ds} - u_{ds} i_{qs}) \end{cases} \tag{14}$$

$$\begin{cases} P_r = \frac{3}{2}(u_{dr} i_{dr} + u_{qr} i_{qr}) \\ Q_r = \frac{3}{2}(u_{qr} i_{dr} - u_{dr} i_{qr}) \end{cases} \tag{15}$$

where P_s, Q_s present stator active and reactive power, respectively. P_r, Q_r presents rotor active and reactive power, respectively. T_{em} is the electromagnetic torque.

The fundamental torque expression is expressed by Equation (16):

$$T_{em} - T_{load} = J \frac{d\omega_m}{dt} \tag{16}$$

With J representing the inertia of the rotor, T_{load} the load torque applied to the shaft and ω_m the rotor speed.

4. MPPT Control

WT consist of four control regions, the wind turbine speed as a function of wind speed is shown in Figure 4. These operating zones can be expresses as follows:

- Zone-1: In this region, speed is very low as the WT cannot generate the power.
- Zone-2: This control zone tracks the restoration of maximum power limited by minimum wind speed to the rated value.
- Zone-3: In this region, the WT operates with rated maximum speed. However, power output is not maximum.
- Zone-4: In this zone, the WT generates maximum rated power. Beyond maximum allowable wind speed, protection devices get activated to avoid failure or damage of WT.

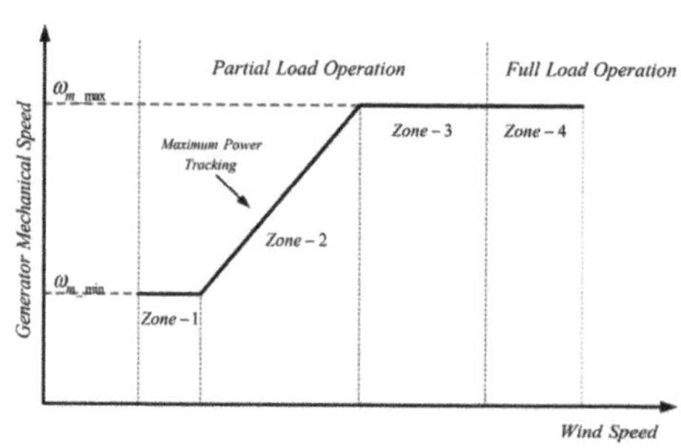

Figure 4. WT control regions.

The main objective of the MPPT controller is to deliver optimal power by DFIG WECS. Here MPPT is used to control the RSC by selecting optimum torque reference for the generator. C_p must be maintained to its $C_{p\,max}$ value to reach the optimum value of torque given by the Equation (18). Figure 5 illustrate the control region range.

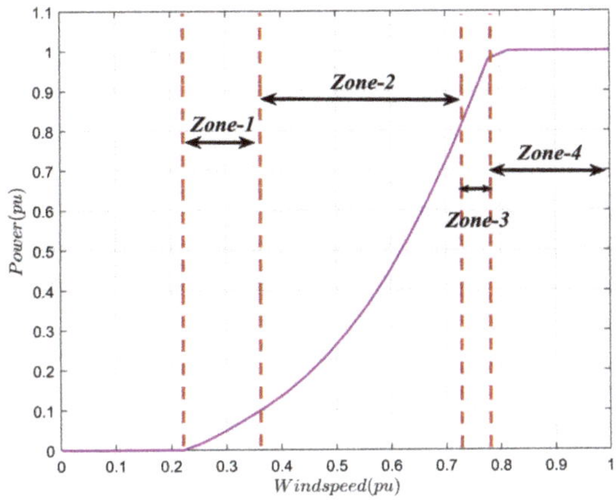

Figure 5. Control region range.

In several studies, different methods have been proposed for wind turbine power extraction [66]. When the WT operate on MPPT, expressions are given by Equation (17):

$$\begin{cases} \lambda_{opt} = \frac{R\omega_t}{V_v} \\ C_p = C_{pmax} \end{cases} \tag{17}$$

The torque is expressed by Equation (18):

$$T_t = \frac{1}{2}\rho\pi\frac{R^5}{\lambda_{opt}^3}C_{pmax}\omega_t^2 = K_{opt}\omega_t^2 \tag{18}$$

where

$$K_{opt} = \frac{1}{2}\rho\pi\frac{R^5}{\lambda_{opt}^3}C_{pmax},\ \beta_{opt} = 0^\circ,\ \omega_m = N\omega_t$$

ω_t is the speed of rotation of the turbine, ω_m is the mechanical speed and N is the multiplier coefficient. Proposed MPPT is presented in Figure 6.

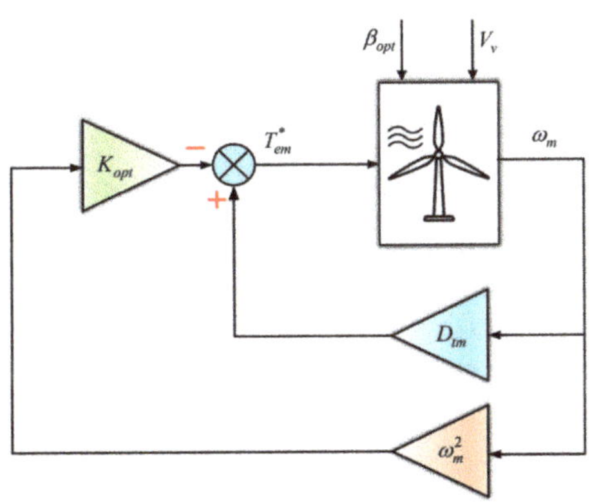

Figure 6. Indirect speed MPPT control.

5. Control Strategy for RSC and GSC

The three-phase grid voltage supplies the DFIG stator winding at constant magnitude and frequency. In contrast, the rotor is supplied at different magnitudes and frequencies by RSC to attain different operating conditions of DFIG. The operating point of the machine decides the power flow through the rotor and grid. The three operating modes of DFIG depend on the speed given in Equation (20):

$$\begin{cases} \omega_s = \omega_r + \omega_m \\ s = \frac{\omega_s - \omega_m}{\omega_s} \end{cases} \tag{19}$$

$$\begin{cases} \omega_m < \omega_s \Rightarrow \omega_s > 0 \Rightarrow s > 0 \Rightarrow \textit{Subsynchronous operation} \\ \omega_m > \omega_s \Rightarrow \omega_r < 0 \Rightarrow s < 0 \Rightarrow \textit{Hypersynchronous operation} \\ \omega_m = \omega_s \Rightarrow \omega_r = 0 \Rightarrow s = 0 \Rightarrow \textit{Synchronous operation} \end{cases} \tag{20}$$

In vector control, a strategy is utilized in *dq* frame with DFIG. This naturally decouples the *d* and *q* quantities. Decoupling makes the DFIG operate as a DC motor. The stator flux vector is aligned along the d axis, as shown in Figure 7.

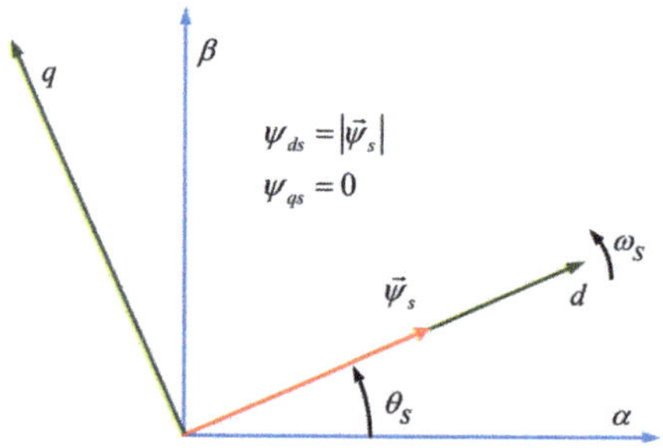

Figure 7. The stator flux vector aligned with d axis component.

5.1. Rotor Side Control

The RSC applies the voltage at rotor winding. Substituting Equations (11) and (12) into Equation (10) gives voltage equations in *dq* frame, considering $\psi_{qs} = 0$ yields the following voltage expressions:

$$\begin{cases} u_{dr} = R_r i_{dr} + \sigma L_r \frac{di_{dr}}{dt} - \omega_r \sigma L_r i_{qr} + \frac{L_m}{L_s}\frac{d\vec{\psi}_s}{dt} \\ u_{qr} = R_r i_{qr} + \sigma L_r \frac{di_{qr}}{dt} + \omega_r \sigma L_r i_{dr} + \omega_r \frac{L_m}{L_s}\frac{d\vec{\psi}_s}{dt} \end{cases} \tag{21}$$

where $\sigma = 1 - L_m^2 / L_s L_r$. Due to the fixed grid quantities $d\vec{\psi}_s/dt$ near to zero, the stator winding resistance drop can be neglected, and the stator flux can be treated as constant. It is evident from Equation (21) that the regulators can be employed to control the *dq* component of rotor current. REG-1 represents the reactive power proportional–integral (PI) regulator. For both *d* and *q* current loops, equal PI regulators are chosen and presented as REG-2 and REG-3, respectively. The gain parameter of the regulator is tuned by considering actual values only. Table 1 represents the value for gain parameters for all three of the regulators. In order to transform the rotor voltage and current into *dq* components using *abc–dq* transform, it is compulsory to apply the control strategy on the *dq* components. Θ_s is obtained by first estimating the stator voltage vector and subtracting angle $\pi/2$. Grid synchronization is achieved by the phase-locked loop (PLL), which in turn will mitigate minor disturbances. The "u", that is 1/3, defines the turn ratio for stator-rotor.

Table 1. The rotor-side control gain parameter of REG-1, REG-2 and REG-3.

Gains	REG-1	REG-2	REG-3
Proportional	10,160	0.5771	0.5771
Integral	406,400	491.5995	491.5995

Expressions for torque in dq frame may be found in Equation (22):

$$T_{em} = \frac{3}{2}p\frac{L_m}{L_s}(\psi_{qs}i_{dr} - \psi_{ds}i_{qr}) = \frac{3}{2}p\frac{L_m}{L_s}\psi_{ds}i_{qr} \Rightarrow T_{em} = K_t i_{qr} \tag{22}$$

The equations for stator reactive power in dq frame may be found in Equation (23):

$$Q_s = \frac{3}{2}(u_{qs}i_{ds} - u_{ds}i_{qs}) = -\frac{3}{2}\omega_s\frac{L_m}{L_s}\vec{\psi}_s\left(i_{dr} - \frac{\vec{\psi}_s}{L_m}\right) \Rightarrow Q_s = K_q\left(i_{dr} - \frac{\vec{\psi}_s}{L_m}\right) \tag{23}$$

Equation (22) reveal that the i_{qr} is proportional to the T_{em}, so torque can be controlled with i_{qr}. Expression in Equation (23) reveals that the Q_s is controlled using the i_{dr}. The RSC control strategy block diagram is presented in Figure 8, which reveals all control loops and MPPT.

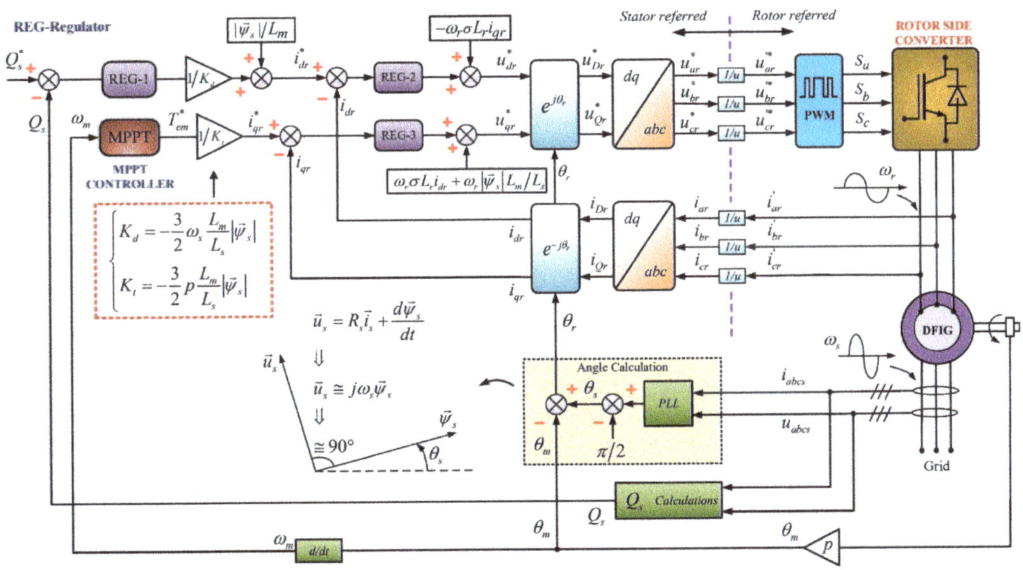

Figure 8. RSC control strategy of the DFIG.

5.2. Grid Side Control

The system configured by the GSC, filter, and grid voltage can be represented as shown in Figure 9. GSC control strategy is used to control the power flow of the DFIG.

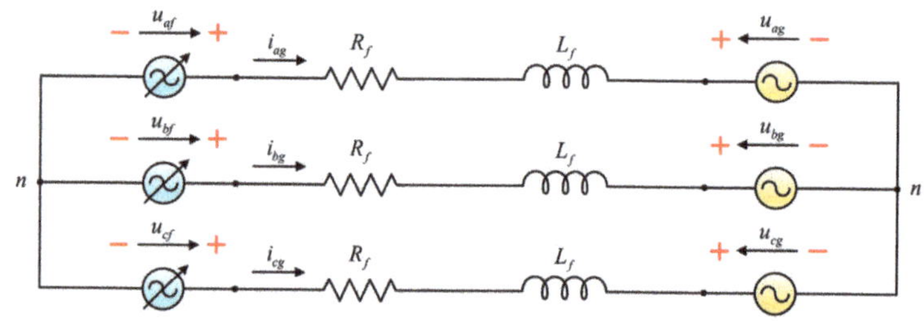

Figure 9. Simplified representation of the three-phase grid system.

Two critical components to consider when controlling the power flow are DC link voltage V_{bus} and reactive power Q_g exchange with grid

Figure 10 presents the dq model of the grid side system in a stationary frame. Figure 11 shows the direct component of voltage vector oriented along the ω_s, considering Equation (24). The dq component of the filter voltage can be expressed by Equation (25) and the expression of the grid exchange active and reactive power in Equation (26).

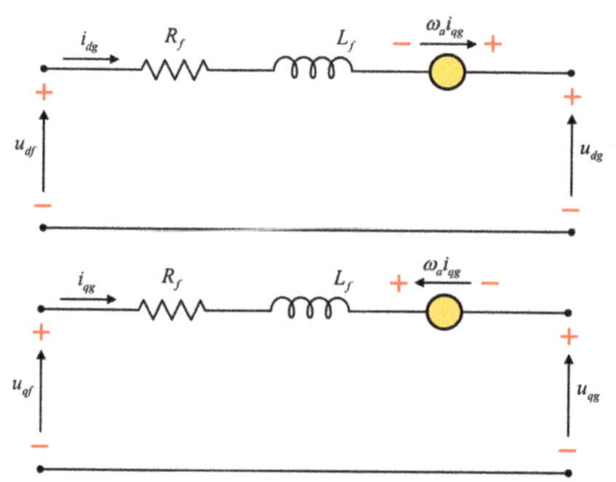

Figure 10. The schematic representation of dq model of the grid side system in the stationary frame.

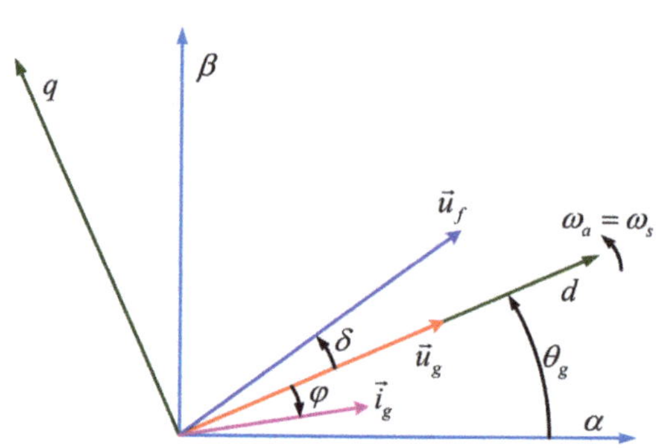

Figure 11. Grid voltage vector oriented along with ω_s.

$$\begin{cases} u_{dg} = \left| \overrightarrow{u}_g \right| \\ u_{qg} = 0 \\ \omega_a = \omega_s \\ \theta = \omega_a t \Rightarrow \theta = \theta_g = \omega_s t \end{cases} \quad (24)$$

$$\begin{cases} u_{df} = R_f i_{dg} + L_f \frac{di_{dg}}{dt} + u_{dg} - \omega_s L_f i_{qg} \\ u_{qf} = R_f i_{qg} + L_f \frac{di_{qg}}{dt} + \omega_s L_f i_{dg} \end{cases} \quad (25)$$

$$\begin{cases} P_g = \frac{3}{2}\left(u_{dg} i_{dg} + u_{qg} i_{qg}\right) \Rightarrow P_g = \frac{3}{2} u_{dg} i_{dg} \\ Q_g = \frac{3}{2}\left(u_{qg} i_{dg} - u_{dg} i_{qg}\right) \Rightarrow Q_g = -\frac{3}{2} u_{dg} i_{qg} \end{cases} \quad (26)$$

Equation (26) reveals that the i_{dg} current component controls the P_g, while the i_{qg} current component controls the Q_g value. Figure 12 shows the block diagram of GSC control. A capacitor forms the DC link; active power flows through RSC-Capacitor-GSC to the grid. Therefore, maintaining V_{bus} to a constant value will ensure both RSC and GSC work properly during active power flow. In the same manner, reactive power flow in the grid is ensured. From reference V_{bus} and Q_g, this generates pulses for the GSC switches S_{ag}, S_{bg} and S_{cg}. Vector control strategy for the GSC is shown in Figure 13. The gain parameters for all three regulators in grid-side control are presented in Table 2.

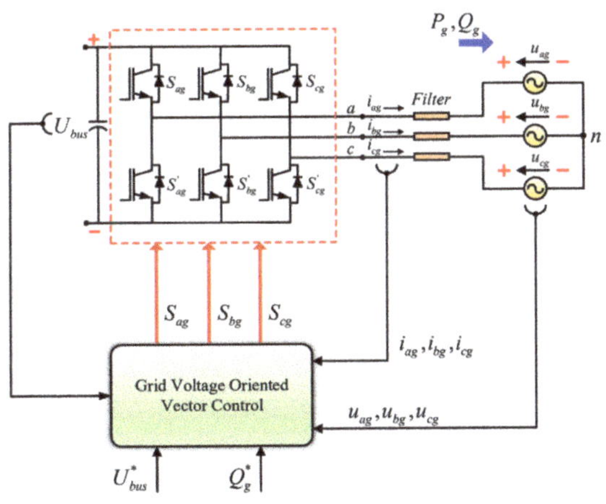

Figure 12. Block diagram of grid-side system.

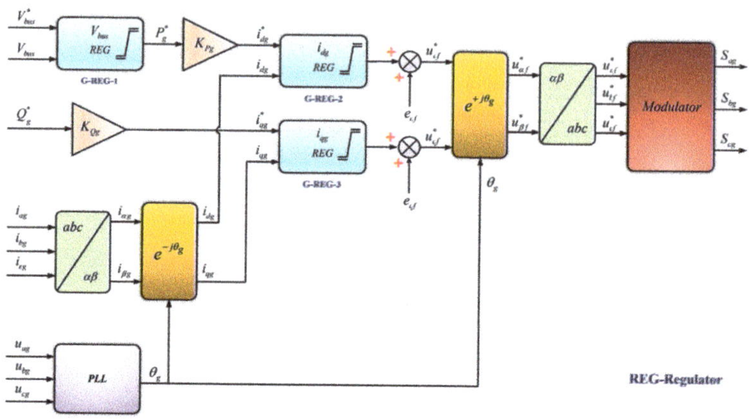

Figure 13. Block diagram of GSC vector control strategy.

Table 2. The gain parameters of PI regulators in grid-side control.

Gains	G-REG-1	G-REG-2	G-REG-3
Proportional	1000	0.3016	0.3016
Integral	300,000	56.8489	56.8489

For improved performance and dynamic response, the coupling term at each current controller output is added by Equation (27). The constant term K_{Pg} and K_{Qg} can be derived from Equation (26) presented in Equation (28).

$$\begin{cases} e_{df} = -\omega_s L_f i_{qg} \\ e_{qf} = \omega_s L_f i_{dg} \end{cases} \tag{27}$$

$$\begin{cases} K_{Pg} = \frac{1}{\frac{3}{2}u_{dg}} \\ K_{Pg} = \frac{1}{-\frac{3}{2}u_{dg}} \end{cases} \tag{28}$$

6. Simulation Results

The proposed DFIG modeling and control of RSC and GCS are implemented and simulated in MATLAB/Simulink environments. In this section, the performance of the proposed system is analyzed during wind speed variation. The simulation parameters of the system are presented in Table 3.

Table 3. Model Simulation Parameters.

Parameter	Parameter Value	Parameter	Parameter Value
Nominal wind speed	11 m/s	Frequency	50 Hz
Air density	1.225 kg/m^3	Rated torque	12,732 N·m
Tip speed ratio	7	Pole pair	2
Pitch angle	0°	Inertia	127 kg·m^2
Power coefficient	0.4411	Gear ratio	100
Nominal Power	2 MW	Radius of turbine	42 m

In this work, β is set to zero and designed for the rated wind speed of 11 m/s. The simulated power characteristics at $\beta = 0^{\circ}$ and different wind speed is presented in Figure 3.$C_p - \lambda_i$ characteristics at different value of β is presented in Figure 14. The design shows that the maximum value of $C_{p\,\max}$ is 0.4411 and corresponding λ is 7 at $\beta = 0^{\circ}$ is shown in Figure 15. This value of $C_{p\,\max}$ and λ is optimum value for capturing peak power from the available wind power.

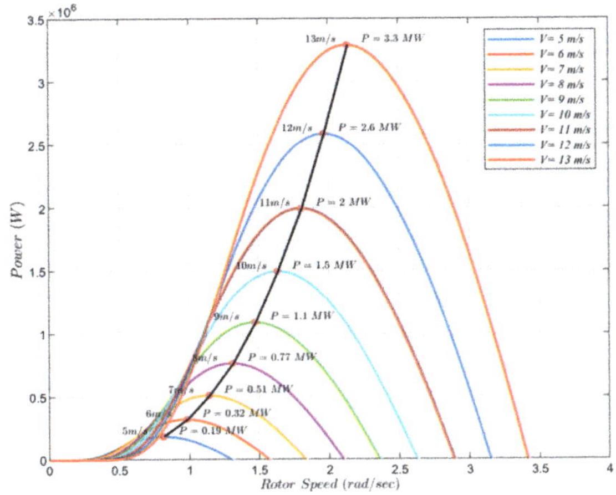

Figure 14. Power and rotor speed characteristics of WT at different wind speed.

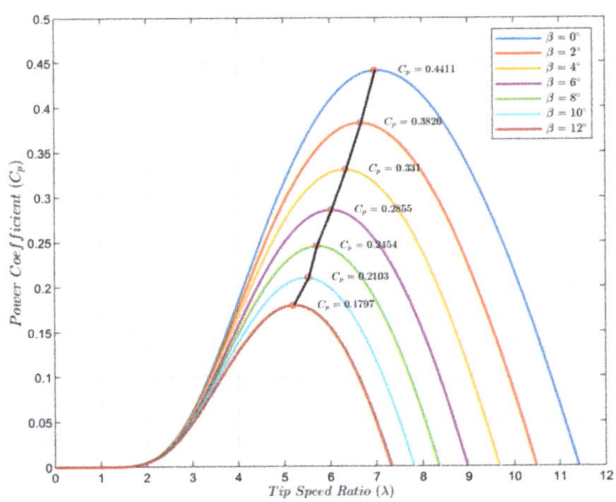

Figure 15. Characteristics with variation in pitch angle (β).

Figure 16 represents the model of DFIG in MATLAB/Simulink environment. Figures 17 and 18 show the Simulink modeling of the RSC control and GSC control.

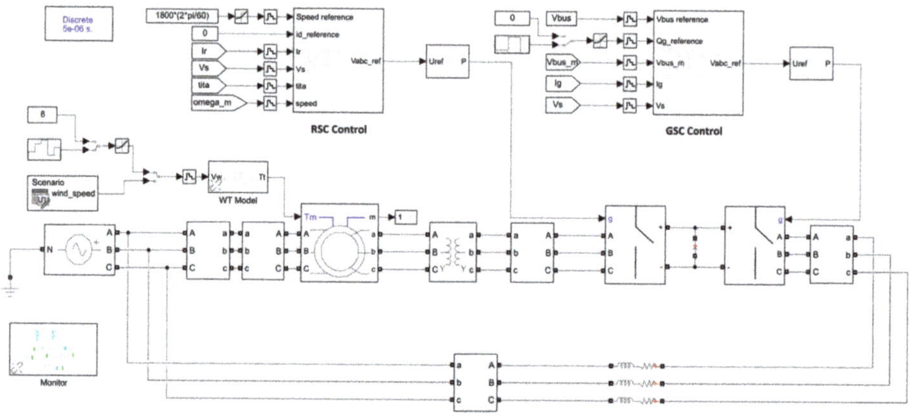

Figure 16. Simulink model of grid-connected DFIG.

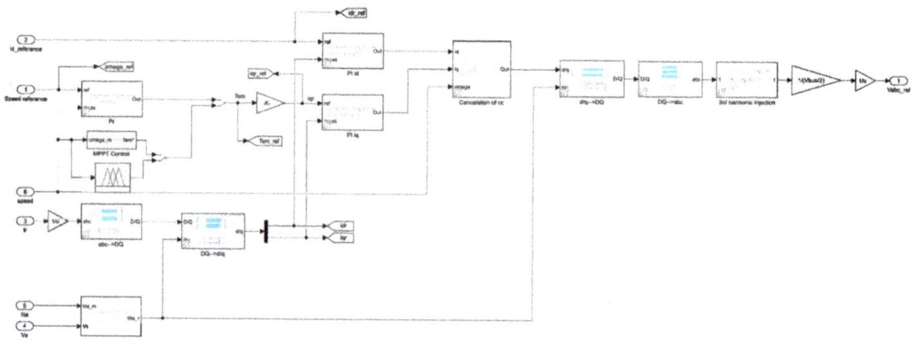

Figure 17. Rotor side control block.

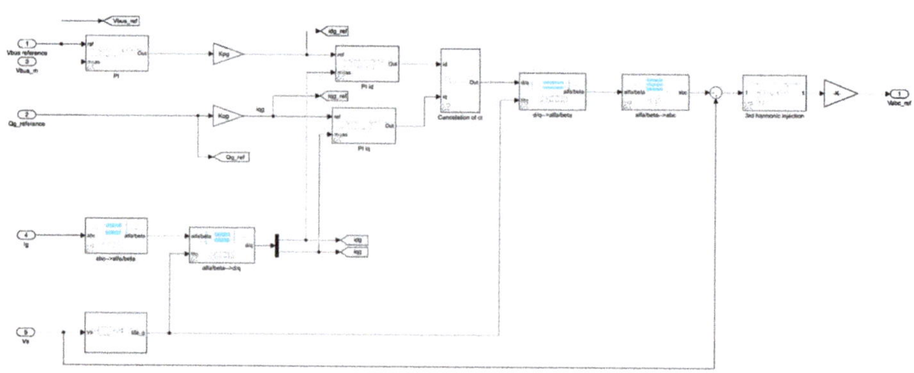

Figure 18. Grid side control block.

The wind speed profile applied to the system is depicted in Figure 19. The wind speed profile has a wide speed range variation between 7 to 12 m/s. In this simulation study the ramp increase and ramp decrease of wind speed is considered. At the start of the simulation a wind speed of 7 m/s is considered, and the system starts simulating from stand steel condition. Figure 20 presents the simulation response of theoretical and actual rotor speed tracking related to the wind speed variation. It can be seen that the proposed control scheme performs well in terms of the tracking of speed.

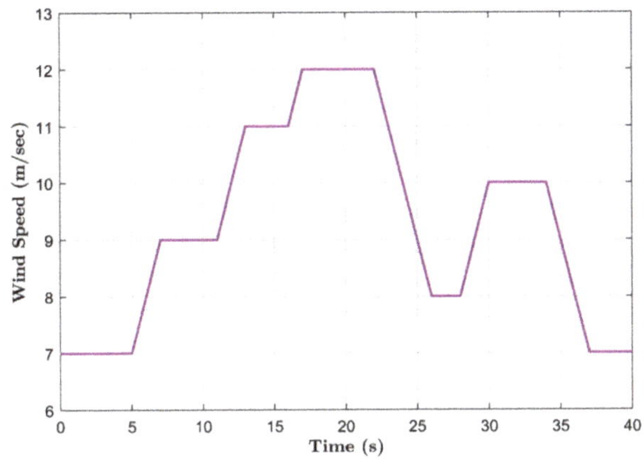

Figure 19. Wind speed variation input to the system.

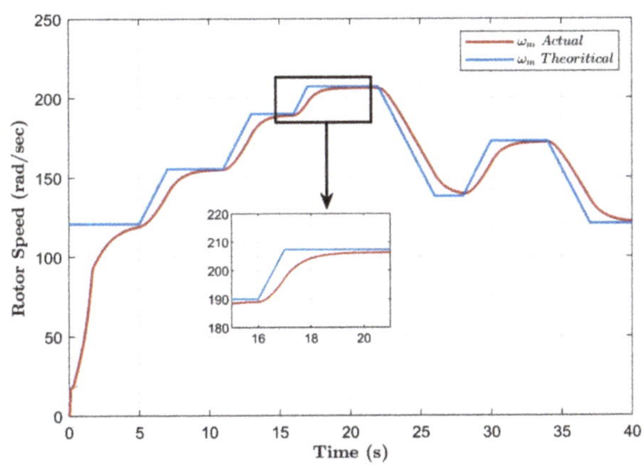

Figure 20. Response of actual and theoretical rotor speed.

Figure 21 shows the variation in the power coefficient (C_p) and tip speed ratio (λ). The blue line represents the C_p and the orange line represents the λ. According to the Betz limit, the maximum theoretical value C_p is expressed in Equation (6) and the optimal value for the proposed system is 0.4411 at zero pitch angle as represented in Figure 15. During the complete simulation, the proposed controllers keep this near to this optimal value which shows a minimal variation C_p between 0.42 to 0.45 due to the excellent perturbation of the MPPT controller. The tracking performance of the electromagnetic torque is presented in Figure 22, which shows the torque reference produced by the MPPT controller and the actual torque of DFIG.

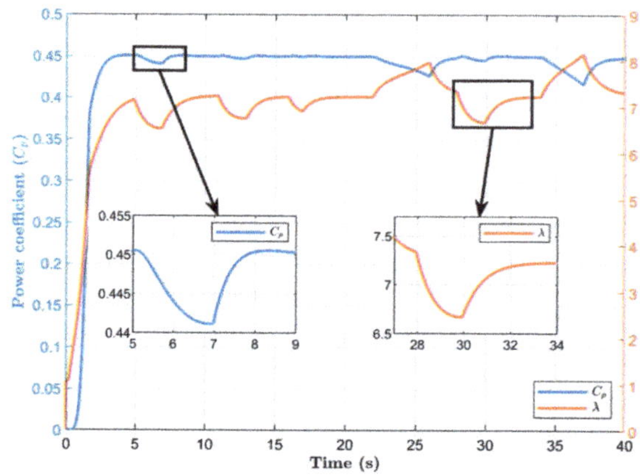

Figure 21. Response of power coefficient and tip speed ratio variation.

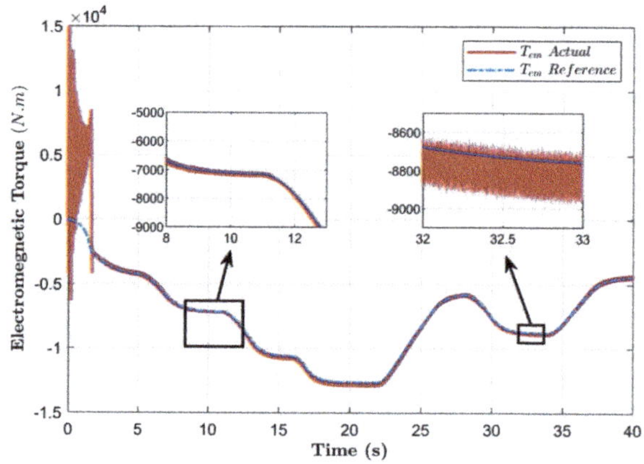

Figure 22. Reference and actual electromagnetic torque tracking.

The stator current and rotor current response during the simulation is presented in Figures 23 and 24, respectively. Figure 25 shows that the single-phase stator voltage and current are balanced and sinusoidal according to the IEEE 519 standard.

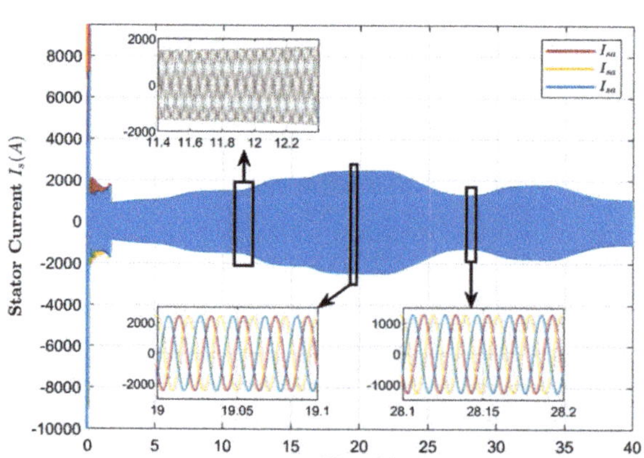

Figure 23. DFIG stator current variation.

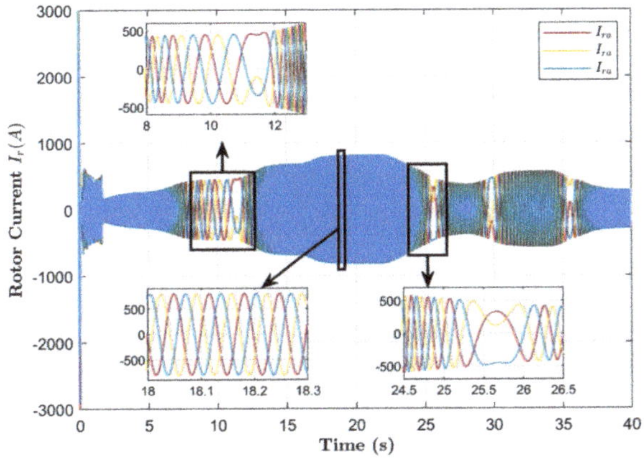

Figure 24. DFIG rotor current variation.

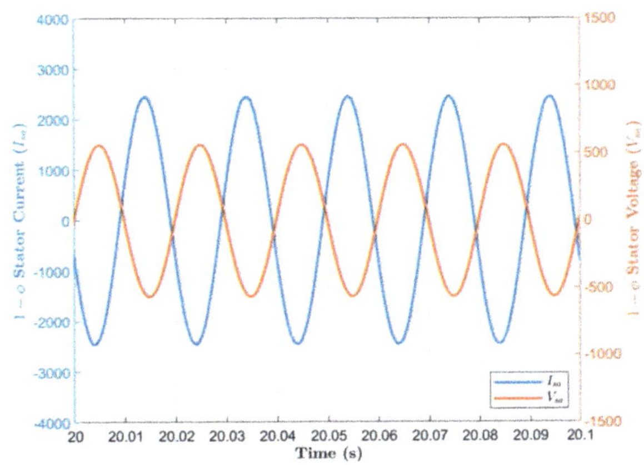

Figure 25. Single phase voltage and current of DFIG stator.

Figure 26 presents the stator active power performance according to wind speed variation. MPPT controller shows excellent dynamic performance during the wind speed variation. As seen in Figure 26, power output at a wind speed of 12 m/s is 1.978 MW. Grid side reactive power tracking performance according to reactive reference power is shown in Figure 27. At t = 10 s, reactive power reference is changed from 0 to -4×10^5, the controller performance shows that the reactive power follows the reference without significant overshoot.

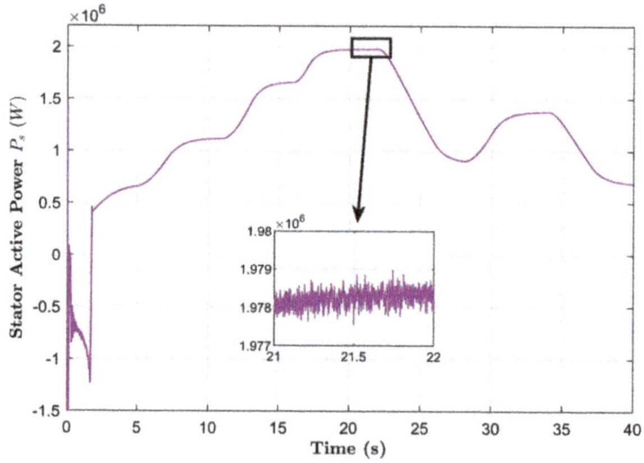

Figure 26. Stator active power variation.

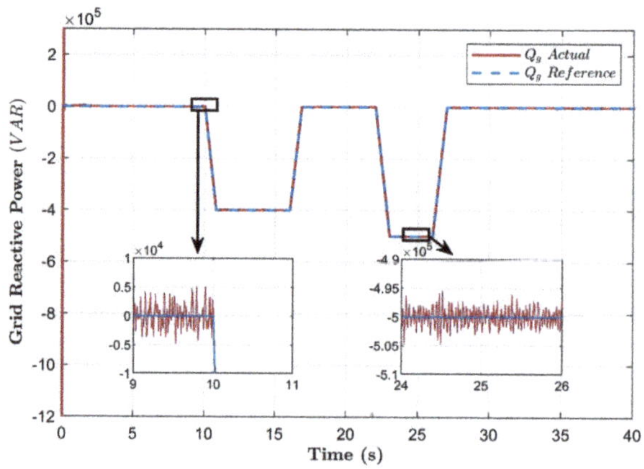

Figure 27. DFIG grid side reactive power variation.

The quadrature and direct current components are responsible for the independent control over active and reactive power flow. Figure 28 presents the reference tracking of the quadrature component of the rotor current. Figure 29 shows the quadrature current and direct current component of the grid current. The yellow line shows the reference quadrature grid current and the red line presents the actual quadrature grid current. The purple line shows the reference direct grid current and the green line presents the actual direct grid current. During the simulation references for the direct and quadrature grid current are changed. The proposed controllers show good tracking under reference variation.

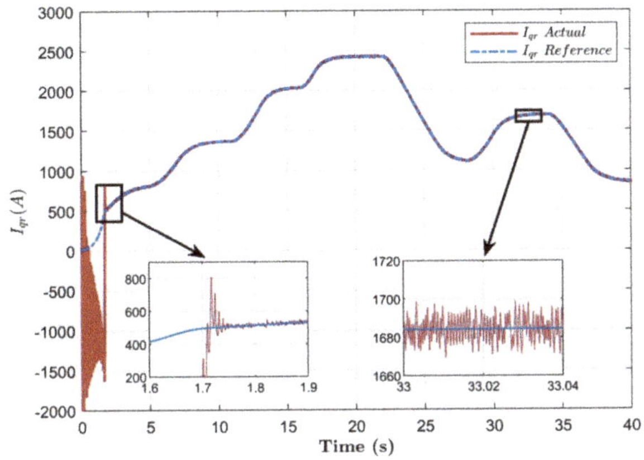

Figure 28. Reference and actual quadrature rotor current tracking performance.

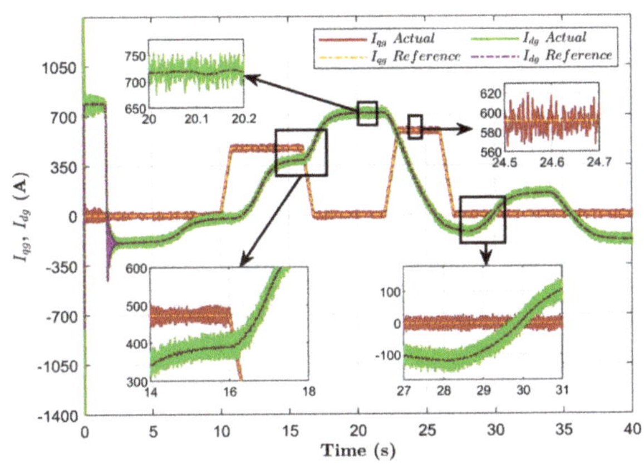

Figure 29. Tracking performance of direct and quadrature current component of GSC.

The reference voltage for the DC link is set to the value of 1150 V. The actual DC link voltage follows the reference very well without the overshooting, and deviation from the reference value is only about 1 V even if the wind speed is changed as is shown in Figure 30. The power flow in the DC link is bidirectional.

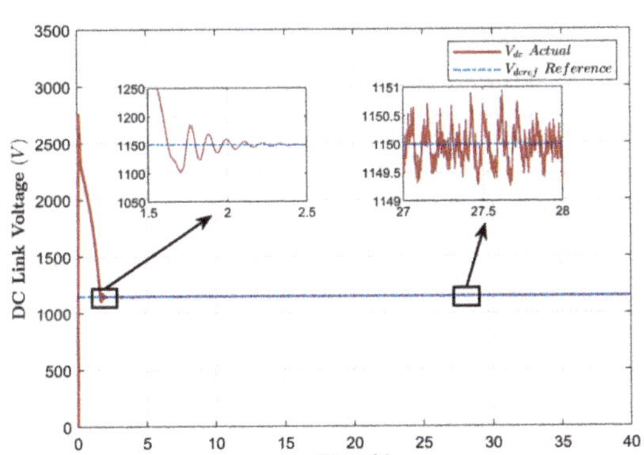

Figure 30. Tracking performance of DC link voltage.

The THD analysis for the stator current from the frequency range of 0 Hz to 1 kHz for three cycles with the fundamental frequency of 50 Hz is shown in Figure 31. This clearly shows that when the generator runs at its rated speed, the THD content in the stator current is only 0.73%.

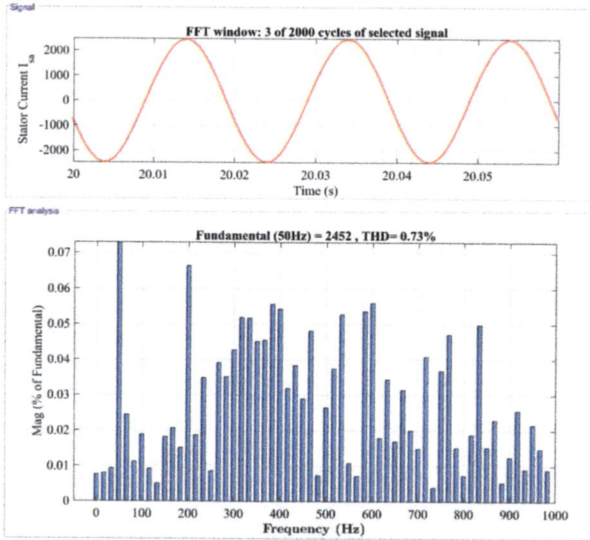

Figure 31. THD performance of single-phase stator current.

Figure 32 shows the THD analysis for the grid current for three cycles, with a frequency range of 0 Hz to 1 kHz with the fundamental frequency of 50 Hz. As seen from Figure 32, the THD content in grid current is only 3.86%, according to IEEE standards.

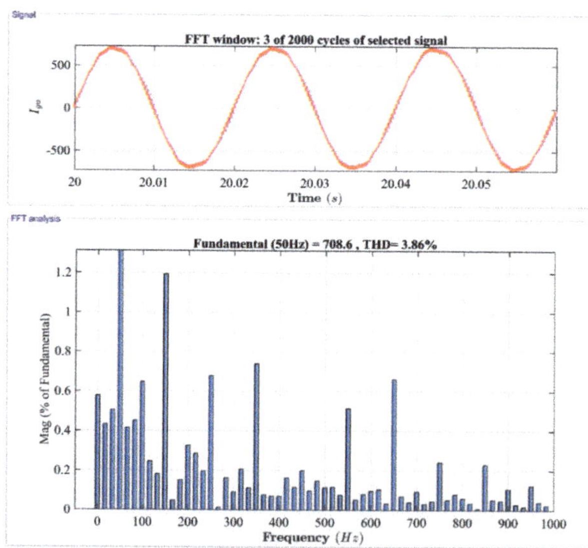

Figure 32. THD performance of single-phase grid current.

From the simulation analysis and all the figures, we conclude that MPPT control shows good power tracking performance under different wind speed conditions. The vector control strategy for RSC and GSC with PI regulators shows good dynamic performance and reference tracking without overshooting. The proposed controls ensure good dynamic and transient performance with maximum power extraction from the wind.

7. Conclusions

The modeling and control of grid-connected DFIG-based WECS has been proposed in this study. At first, the mathematical model of turbine and DFIG are proposed. The proposed MPPT controller works efficiently with wind speed variation and tracks the power very well, even in low wind conditions to the wide wind speed variation. The vector control strategy for the RSC and GSC have been proposed for the power flow control between the grid and the DFIG system. The stator voltage and current waveform are balanced and sinusoidal. The simulation study shows that the proposed system with a control strategy works satisfactorily and gives good performance under wind speed variation. THD analysis shows that the stator current and grid current THD is according to IEEE 519 standard. Moreover, results show fast response time, no overshoot, and robustness of the proposed system. This research may be further explored, considering the variable pitch angle along with multi-input variables. Performance may also be contrasted with other controller types based on artificial intelligence, model reference adaptive system that may be substituted in the RSC and GSC schemes.

Author Contributions: Conceptualization, A.A.C., P.C. and A.N.V.; methodology, P.C. and V.B.; investigation, A.A.C., T.C. and S.G.; formal analysis, V.B., G.S. and A.N.V.; validation, V.B., G.S. and A.K.; resources and data curation, A.A.C. and V.B.; writing—original draft preparation, A.A.C., T.C. and S.G.; writing review and editing, P.C. and V.B.; visualization, A.A.C., V.B. and TC.; supervision, P.C., V.B. and G.S.; project administration, P.C.; funding acquisition, V.B., G.S., A.N.V. and A.K. All authors have read and agreed to the published version of the manuscript.

Funding: This research received no external funding.

Informed Consent Statement: Not applicable.

Data Availability Statement: Data are available on request to abrar0613@gmail.com.

Conflicts of Interest: The authors declare no conflict of interest.

References

1. Chhipa, A.A.; Kumar, V.; Vyas, S.; Joshi, R.R. MPPT Optimisation Techniques and Power Electronics for Renewable Energy Systems: Wind and Solar Energy Systems. *Int. J. Swarm Intell.* **2021**, *1*, 1. [CrossRef]
2. Qazi, A.; Hussain, F.; Rahim, N.A.B.D.; Hardaker, G.; Alghazzawi, D.; Shaban, K.; Haruna, K. Towards Sustainable Energy: A Systematic Review of Renewable Energy Sources, Technologies, and Public Opinions. *IEEE Access* **2019**, *7*, 63837–63851. [CrossRef]
3. Prasad, R.M.; Mulla, M.A. Mathematical Modeling and Position-Sensorless Algorithm for Stator-Side Field-Oriented Control of Rotor-Tied DFIG in Rotor Flux Reference Frame. *IEEE Trans. Energy Convers.* **2020**, *35*, 631–639. [CrossRef]
4. Hu, J.; Nian, H.; Xu, H.; He, Y. Dynamic Modeling and Improved Control of DFIG under Distorted Grid Voltage Conditions. *IEEE Trans. Energy Convers.* **2011**, *26*, 163–175. [CrossRef]
5. Hosseini, S.H.; Sharifian, M.B.B.; Shahnia, F. Dynamic Performance of Double Fed Induction Generator for Wind Turbines. In Proceedings of the ICEMS 2005 Eighth International Conference on Electrical Machines and Systems, Nanjing, China, 27–29 September 2005; Volume 2, pp. 1261–1266. [CrossRef]
6. Balat, M.; Balat, H. Biogas as a Renewable Energy Sourcea Review. *Energy Sources Part A Recovery Util. Environ. Eff.* **2009**, *31*, 1280–1293. [CrossRef]
7. Chhipa, A.A.; Vyas, S.; Kumar, V.; Joshi, R.R. Role of Power Electronics and Optimization Techniques in Renewable Energy Systems. In *Intelligent Algorithms for Analysis and Control of Dynamical Systems. Algorithms for Intelligent Systems*; Springer: Singapore, 2021; pp. 167–175. [CrossRef]
8. Hansen, A.D.; Sørensen, P.; Lov, F.; Blaabjerg, F. Control of Variable Speed Wind Turbines with Doubly-Fed Induction Generators. *Wind Eng.* **2004**, *28*, 411–432. [CrossRef]
9. Tanvir, A.A.; Merabet, A.; Beguenane, R. Real-Time Control of Active and Reactive Power for Doubly Fed Induction Generator (DFIG)-Based Wind Energy Conversion System. *Energies* **2015**, *8*, 10389–10408. [CrossRef]

10. Poitiers, F.; Bouaouiche, T.; Machmoum, M. Advanced Control of a Doubly-Fed Induction Generator for Wind Energy Conversion. *Electr. Power Syst. Res.* **2009**, *79*, 1085–1096. [CrossRef]
11. Rezaei, E.; Tabesh, A.; Ebrahimi, M. Dynamic Model and Control of DFIG Wind Energy Systems Based on Power Transfer Matrix. *IEEE Trans. Power Deliv.* **2012**, *27*, 1485–1493. [CrossRef]
12. Zou, Y.; Elbuluk, M.; Sozer, Y. A Complete Modeling and Simulation of Induction Generator Wind Power Systems. In Proceedings of the Conference Record-IAS Annual Meeting (IEEE Industry Applications Society), Houston, TX, USA, 3–7 October 2010. [CrossRef]
13. Idrissi, I.; Chafouk, H.; El Bachtiri, R.; Khanfara, M. Modeling and Simulation of the Variable Speed Wind Turbine Based on a Doubly Fed Induction Generator. In *Gas Turbines-Control, Diagnostics, Simulation, and Measurements*; IntechOpen: London, UK, 2019. [CrossRef]
14. Ullah, N.R.; Thiringer, T. Variable Speed Wind Turbines for Power System Stability Enhancement. *IEEE Trans. Energy Convers.* **2007**, *22*, 52–60. [CrossRef]
15. Chhipa, A.A.; Kumar, V.; Joshi, R.R.; Chakrabarti, P.; Jasinski, M.; Burgio, A.; Leonowicz, Z.; Jasinska, E.; Soni, R.; Chakrabarti, T. Adaptive Neuro-Fuzzy Inference System-Based Maximum Power Tracking Controller for Variable Speed WECS. *Energies* **2021**, *14*, 6275. [CrossRef]
16. Rao, K.R. Wind Energy: Technical Considerations–Contents. In *Wind Energy for Power Generation*; Springer: Cham, Switzerland, 2019; pp. 1–426. [CrossRef]
17. Xiahou, K.; Liu, Y.; Wang, L.; Li, M.S.; Wu, Q.H. Control of DFIG's Rotor-Side Converter with Decoupling of Current Loops Using Observer-Based Fractional-Order Sliding-Mode Regulators. *IEEE Access* **2019**, *7*, 163412–163420. [CrossRef]
18. Yao, J.; Li, H.; Chen, Z.; Xia, X.; Chen, X.; Li, Q.; Liao, Y. Enhanced Control of a DFIG-Based Wind-Power Generation System with Series Grid-Side Converter under Unbalanced Grid Voltage Conditions. *IEEE Trans. Power Electron.* **2013**, *28*, 3167–3181. [CrossRef]
19. Rajendran, M.; Kumar, L.A. Modeling and Simulation of a Dfig-Based Wind Energy System. In *Lecture Notes in Electrical Engineering*; Springer: Singapore, 2020; Volume 687, pp. 31–49. [CrossRef]
20. Nian, H.; Xu, Y.; Chen, L.; Zhu, M. Modeling and Analysis of DC-Link Dynamics in DFIG System with an Indicator Function. *IEEE Access* **2019**, *7*, 125401–125412. [CrossRef]
21. Karabacak, K.; Cetin, N. Artificial Neural Networks for Controlling Wind-PV Power Systems: A Review. *Renew. Sustain. Energy Rev.* **2014**, *29*, 804–827. [CrossRef]
22. Marugán, A.P.; Márquez, F.P.G.; Perez, J.M.P.; Ruiz-Hernández, D. A Survey of Artificial Neural Network in Wind Energy Systems. *Appl. Energy* **2018**, *228*, 1822–1836. [CrossRef]
23. Gayatri, M.T.L.; Parimi, A.M.; Pavan Kumar, A.V. A Review of Reactive Power Compensation Techniques in Microgrids. *Renew. Sustain. Energy Rev.* **2018**, *81*, 1030–1036. [CrossRef]
24. Geng, H.; Liu, C.; Yang, G. LVRT Capability of DFIG-Based WECS under Asymmetrical Grid Fault Condition. *IEEE Trans. Ind. Electron.* **2013**, *60*, 2495–2509. [CrossRef]
25. Singh, G.K. Self-Excited Induction Generator Research—A Survey. *Electr. Power Syst. Res.* **2004**, *69*, 107–114. [CrossRef]
26. Singh, B.; Murthy, S.S.; Gupta, S. A Voltage and Frequency Controller for Self-Excited Induction Generators. *Electr. Power Compon. Syst.* **2006**, *34*, 141–157. [CrossRef]
27. McPherson, G. Analysis of the Isolated Induction Generator. *IEEE Trans. Power Appar. Syst.* **1983**, *8*, 2793–2798. [CrossRef]
28. Casielles, P.G.; Sanz, J.; Pascual, J. Control System for a Wind Turbine Driven Self Excited Asynchronous Generator. In Proceedings of the Electrotechnical Conference Integrating Research, Industry and Education in Energy and Communication Engineering, Lisbon, Portugal, 11–13 April 1989; pp. 95–98. [CrossRef]
29. Kasal, G.K.; Singh, B. Voltage and Frequency Controllers for an Asynchronous Generator-Based Isolated Wind Energy Conversion System. *IEEE Trans. Energy Convers.* **2011**, *26*, 402–416. [CrossRef]
30. Peña, R.; Cárdenas, R.; Escobar, E.; Clare, J.; Wheeler, P. Control System for Unbalanced Operation of Stand-Alone Doubly Fed Induction Generators. *IEEE Trans. Energy Convers.* **2007**, *22*, 544–545. [CrossRef]
31. Jain, A.K.; Ranganathan, V.T. Wound Rotor Induction Generator with Sensorless Control and Integrated Active Filter for Feeding Nonlinear Loads in a Stand-Alone Grid. *IEEE Trans. Ind. Electron.* **2008**, *55*, 218–228. [CrossRef]
32. Goel, P.K.; Singh, B.; Murthy, S.S.; Kishore, N. Parallel Operation of DFIGs in Three-Phase Four-Wire Autonomous Wind Energy Conversion System. *IEEE Trans. Ind. Appl.* **2011**, *47*, 1872–1883. [CrossRef]
33. Watson, D.B.; Arrillaga, J.; Densem, T. Controllable d.c. power supply from wind-driven self-excited induction machines. *Proc. Inst. Electr. Eng.* **1979**, *126*, 1245–1248. [CrossRef]
34. Chen, W.; Liming, W.; Libao, S.; Yixin, N. A Survey on Wind Power Technologies in Power Systems. In Proceedings of the 2007 IEEE Power Engineering Society General Meeting, PES, Tampa, FL, USA, 24–28 June 2007. [CrossRef]
35. Hinrichsen, E.N. Controls for Variable Pitch Wind Turbine Generators. *IEEE Trans. Power Appar. Syst.* **1984**, *4*, 886–892. [CrossRef]
36. Yamamoto, M.; Motoyoshi, O. Active and Reactive Power Control for Doubly-Fed Wound Rotor Induction Generator. *IEEE Trans. Power Electron.* **1991**, *6*, 624–629. [CrossRef]
37. Uctug, M.Y.; Eskandarzadeh, I.; Ince, H. Modelling and Output Power Optimisation of a Wind Turbine Driven Double Output Induction Generator. *IEE Proc. Electr. Power Appl.* **1994**, *141*, 33–38. [CrossRef]

38. Arnaltes Gómez, S.; Rodriguez Amenedo, J.L. Grid Synchronisation of Doubly Fed Induction Generators Using Direct Torque Control. In Proceedings of the IEEE 2002 28th Annual Conference of the Industrial Electronics Society. IECON, Seville, Spain, 5–8 November 2002; Volume 4, pp. 3338–3343. [CrossRef]
39. Tapia, A.; Tapia, G.; Xabier Ostolaza, J.; Sáenz, J.R. Modeling and Control of a Wind Turbine Driven Doubly Fed Induction Generator. *IEEE Trans. Energy Convers.* **2003**, *18*, 194–204. [CrossRef]
40. Jabr, H.M.; Kar, N.C. Adaptive Vector Control for Slip Energy Recovery in Doubly-Fed Wind Driven Induction Generator. In Proceedings of the Canadian Conference on Electrical and Computer Engineering, Saskatoon, SK, Canada, 1–4 May 2005; Volume 2005, pp. 759–762. [CrossRef]
41. Jiao, L.; Ooi, B.T.; Joós, G.; Zhou, F. Doubly-Fed Induction Generator (DFIG) as a Hybrid of Asynchronous and Synchronous Machines. *Electr. Power Syst. Res.* **2005**, *76*, 33–37. [CrossRef]
42. Tremblay, E.; Chandra, A.; Lagacé, P.J. Grid-Side Converter Control of DFIG Wind Turbines to Enhance Power Quality of Distribution Network. In Proceedings of the 2006 IEEE Power Engineering Society General Meeting, PES, Montreal, QC, Canada, 18–22 June 2006. [CrossRef]
43. Chowdhury, B.H.; Chellapilla, S. Double-Fed Induction Generator Control for Variable Speed Wind Power Generation. *Electr. Power Syst. Res.* **2006**, *76*, 786–800. [CrossRef]
44. Datta, R.; Ranganathan, V.T. Decoupled Control of Active and Reactive Power for a Grid-Connected Doubly-Fed Wound Rotor Induction Machine without Position Sensors. In Proceedings of the Conference Record-IAS Annual Meeting (IEEE Industry Applications Society), Phoenix, AZ, USA, 3–7 October 1999; Volume 4, pp. 2623–2630. [CrossRef]
45. Datta, R.; Ranganathan, V.T. Direct Power Control of Grid-Connected Wound Rotor Induction Machine without Rotor Position Sensors. *IEEE Trans. Power Electron.* **2001**, *16*, 390–399. [CrossRef]
46. Ying, L.M.; Cui, X.; Liao, Q.F.; Tang, C.H.; Le, L.C.; Chen, Z. Stator Flux Observation and Speed Estimation of a Doubly Fed Induction Generator. In Proceedings of the 2006 International Conference on Power System Technology, POWERCON2006, Chongqing, China, 22–26 October 2006. [CrossRef]
47. Qiao, W.; Zhou, W.; Aller, J.M.; Harley, R.G. Wind Speed Estimation Based Sensorless Output Maximization Control for a Wind Turbine Driving a DFIG. *IEEE Trans. Power Electron.* **2008**, *23*, 1156–1169. [CrossRef]
48. Hongfei, M.; Yi, W.; Dianguo, X.; Yongqiang, L. Research on AC-DC-AC Converter for Wind Power Doubly-Fed Induction Generator. In Proceedings of the IECON Proceedings (Industrial Electronics Conference), Paris, France, 6–10 November 2006; pp. 2734–2739. [CrossRef]
49. Pena, R.S.; Cardenas, R.J.; Asher, G.M.; Clare, J.C. Vector Controlled Induction Machines for Stand-Alone Wind Energy Applications. In Proceedings of the Conference Record of the 2000 IEEE Industry Applications Conference. Thirty-Fifth IAS Annual Meeting and World Conference on Industrial Applications of Electrical Energy, Rome, Italy, 8–12 October 2000; Volume 3, pp. 1409–1415. [CrossRef]
50. Cárdenas, R.; Peña, R.; Proboste, J.; Asher, G.; Clare, J. MRAS Observer for Sensorless Control of Standalone Doubly Fed Induction Generators. *IEEE Trans. Energy Convers.* **2005**, *20*, 710–718. [CrossRef]
51. Hilloowala, R.M.; Sharaf, A.M. A Rule-Based Fuzzy Logic Controller for a PWM Inverter in a Stand Alone Wind Energy Conversion Scheme. *IEEE Trans. Ind. Appl.* **1996**, *32*, 57–65. [CrossRef]
52. Simões, M.G.; Bose, B.K.; Spiegel, R.J. Design and Performance Evaluation of a Fuzzy-Logic-Based Variable-Speed Wind Generation System. *IEEE Trans. Ind. Appl.* **1997**, *33*, 956–965. [CrossRef]
53. Poddar, G.; Joseph, A.; Unnikrishnan, A.K. Sensorless Variable-Speed Controller for Existing Fixed-Speed Wind Power Generator with Unity-Power-Factor Operation. *IEEE Trans. Ind. Electron.* **2003**, *50*, 1007–1015. [CrossRef]
54. Teodorescu, R.; Blaabjerg, F. Flexible Control of Small Wind Turbines with Grid Failure Detection Operating in Stand-Alone and Grid-Connected Mode. *IEEE Trans. Power Electron.* **2004**, *19*, 1323–1332. [CrossRef]
55. Ahmed, T.; Nishida, K.; Nakaoka, M. Advanced Control of PWM Converter with Variable-Speed Induction Generator. *IEEE Trans. Ind. Appl.* **2006**, *42*, 934–945. [CrossRef]
56. Ahmed, T.; Nishida, K.; Nakaoka, M. Advanced Control for PWM Converter and Variable-Speed Induction Generator. *IET Electr. Power Appl.* **2007**, *1*, 239–247. [CrossRef]
57. Barakati, S.M.; Kazerani, M.; Aplevich, J.D. Maximum Power Tracking Control for a Wind Turbine System Including a Matrix Converter. *IEEE Trans. Energy Convers.* **2009**, *24*, 705–713. [CrossRef]
58. Kazmi, S.M.R.; Goto, H.; Guo, H.J.; Ichinokura, O. A Novel Algorithm for Fast and Efficient Speed-Sensorless Maximum Power Point Tracking in Wind Energy Conversion Systems. *IEEE Trans. Ind. Electron.* **2011**, *58*, 29–36. [CrossRef]
59. Vas, P.; Li, J. Simulation package for vector controlled induction motor drives. In Proceedings of the 1993 Sixth International Conference on Electrical Machines and Drives, Oxford, UK, 8–10 September 1993; pp. 265–270. Available online: https://ieeexplore.ieee.org/document/253528 (accessed on 3 September 2022).
60. Gabriel, R.; Leonhard, W.; Nordby, C.J. Field-Oriented Control of a Standard AC Motor Using Microprocessors. *IEEE Trans. Ind. Appl.* **1980**, *2*, 186–192. [CrossRef]
61. Wu, Y.K.; Yang, W.H. Different Control Strategies on the Rotor Side Converter in DFIG-Based Wind Turbines. *Energy Procedia* **2016**, *100*, 551–555. [CrossRef]
62. Hore, D.; Sarma, R. Neural Network–based Improved Active and Reactive Power Control of Wind-Driven Double Fed Induction Generator under Varying Operating Conditions. *Wind Eng.* **2018**, *42*, 381–396. [CrossRef]

63. Carroll, J.; McDonald, A.; McMillan, D. Reliability Comparison of Wind Turbines with DFIG and PMG Drive Trains. *IEEE Trans. Energy Convers.* **2015**, *30*, 663–670. [CrossRef]
64. Hosseini, S.M.H.; Rezvani, A. Modeling and Simulation to Optimize Direct Power Control of DFIG in Variable-Speed Pumped-Storage Power Plant Using Teaching–learning-Based Optimization Technique. *Soft Comput.* **2020**, *24*, 16895–16915. [CrossRef]
65. Xia, Y.; Chen, Y.; Song, Y.; Strunz, K. Multi-Scale Modeling and Simulation of DFIG-Based Wind Energy Conversion System. *IEEE Trans. Energy Convers.* **2020**, *35*, 560–572. [CrossRef]
66. Abad, G.; López, J.; Rodríguez, M.A.; Marroyo, L.; Iwanski, G. *Doubly Fed Induction Machine: Modeling and Control for Wind Energy Generation*; Institute of Electrical and Electronics Engineers: Piscataway, NJ, USA, 2011. [CrossRef]

energies

Article

Deep Feature Based Siamese Network for Visual Object Tracking

Su-Chang Lim [1], Jun-Ho Huh [2,*] and Jong-Chan Kim [1,*]

1 Department of Computer Engineering, Sunchon National University, Suncheon 57992, Korea
2 Department of Data Science, (National) Korea Maritime and Ocean University, Busan 49112, Korea
* Correspondence: 72networks@kmou.ac.kr (J.-H.H.); seaghost@sunchon.ac.kr (J.-C.K.)

Abstract: One of the most important and challenging research subjects in computer vision is visual object tracking. The information obtained from the first frame consists of limited and insufficient information to represent an object. If prior information about robust representation that can represent an object well is not sufficient, object tracking fails when not robustly responding to changes in features of the target object according to various factors, namely shape, illumination variation, and scene distortion. In this paper, a real-time single object tracking algorithm is proposed based on a Siamese network to solve this problem. For the object feature extraction, we designed a fully convolutional neural network that removes a fully connected layer and configured a convolution block consisting of a bottleneck structure that preserves the information in a previous layer. This network was designed as a Siamese network, while a regional proposal network was combined at the end of the network for object tracking. The ImageNet Large-Scale Visual Recognition Challenge 2017 dataset was used to train the network in the pre-training phase. Then, in the experimental phase, the object tracking benchmark dataset was used to quantitatively evaluate the network. The experimental results revealed that the proposed tracking algorithm produced more competitive results compared to other tracking algorithms.

Keywords: object tracking; convolution neural network; AI; siamese network; image similarity; CUDA; Python; PyTorch; computer vision

Citation: Lim, S.-C.; Huh, J.-H.; Kim, J.-C. Deep Feature Based Siamese Network for Visual Object Tracking. *Energies* **2022**, *15*, 6388. https://doi.org/10.3390/en15176388

Academic Editor: Oscar Barambones

Received: 19 July 2022
Accepted: 18 August 2022
Published: 1 September 2022

1. Introduction

Visual Object Tracking (VOT) is one of the categories in computer vision and plays an important role in various tasks. VOT is widely used in video analysis application programs such as factory automation monitoring, autonomous driving, intruder monitoring, and drone tasks [1–4]. In particular, more recently, VOT analyzes a relationship between similar pixels in different frames. The information of the tracking target is initialized using the information of ground truth of the first frame in the image sequence. The output result of the tracking algorithm provides a boundary box that displays the size and location of the target for a specific frame in the image sequence [5–7].

However, it has a constraint of using only limited information obtained in the first frame. This constraint causes a tracker to drift in the image sequence and tracking failure to increase if prior information about robust representation that can represent an object well is not sufficient [8,9]. Despite there being studies conducted on performance improvements of VOT algorithms, many difficulties still need to be overcome. In the tracking process, situations of failing to robustly respond to changes in features of the target object occur due to various factors, namely shape, illumination variation, and scene distortion that are applied to a video sequence. This results in object tracking failure as a discrepancy between the current target and the original template takes effect [10,11].

A number of various approaches have been proposed to solve these problems in object tracking. A tracker extracts the distinctive robust feature, which is the main key feature, from the target to the extent that the target attributes can be expressed. Using this feature, an appearance is modeled to find the target from the image frame area and remove the external noise elements. To capture a change in the target shape during the tracking process, ultimately an effective feature for object tracking should be designed. Generally, attributes that change in the object's appearance model over time should be reflected or unique features that can represent the object should be extracted.

As methods based on features, there are the correlation filter-based approach and the deep neural network approach. A tracking algorithm based on a correlation filter generates a filter through appearance modeling using the extracted object's features. Filter weight is updated to run training from image samples of the object region that are continuously inputted while tracking progress. This training is performed in the Fourier domain using a fast Fourier transform (FFT) [12,13]. The correlation filter-based method has the advantage of fast operation speed by being computationally efficient. However, its drawback is that image information is represented inaccurately as its information is disturbed, which is caused by a boundary effect [14].

Recent study methods have focused on deep features based on deep learning, shifting from existing hand-crafted methods. A deep feature extraction method offers many advantages of being more apt to encode multi-layer information through multiple layers and being more invariant to changes in target shapes than a hand-crafted feature extraction method. Thus, it is regarded as the key element to overcome the limitation of traditional tracking algorithms. To robustly track an object using deep features, a correlation filter approach is used [15,16]. However, a correlation filter method has to continuously update an appearance model in the tracking process, because even if robust features are added, the original template model gets corrupted by the surrounding background. A deep network provides a generalization capability that captures various features by various training datasets and many parameters in a network. However, a drawback of this is that it cannot adaptively respond to appearance changes, deformation, occlusion, etc.

In this paper, unique features of the target object are extracted using a convolutional neural network (CNN) and then used in the object tracking algorithm. Using high-level features extracted in a CNN, we regard a tracking problem as a similarity comparison problem that finds a specific object within the image. Calculating image similarity entails finding feature compatibility in an image patch and comparing the features of the target object and the features of objects in the image plane. To do this, we created a customized CNN with a Siamese network, which is an architecture with a Y-shaped branch of two same CNNs. This network outputs similar feature information because the same operation is applied to the target object image and an image containing the object using the same weight. We conducted feature extraction and similarity comparison with one-shot learning through this network. A region proposal network (RPN) was used to infer a region where the target object was present from the region with the highest similarity. Using the proposed tracker, deep features of the object were extracted in real time, thereby emphasizing the distinctiveness between objects themselves or between object and background through the feature similarity comparison. Through this, we could improve the tracking algorithm's performance. In particular, we have shown a robust performance in appearance change and distraction factors. There are three contributions to this work.

We analyze features for object tracking using CNNs trained on large image datasets to find important properties. The features of CNNs show better results than traditional tracking algorithms using hand-crafted features, helping to design effective CNN-based trackers.

We propose a method to combine two CNNs with the same structure to form a Siamese network to handle sudden appearance changes and track target objects through similarity comparison between two images.

The proposed tracking algorithm greatly mitigates object drift. We improved the tracking accuracy by introducing the anchor box concept that estimates the object area through similarity comparison between feature maps extracted from CNN. The evaluation of popular tracking benchmarks shows that the proposed method handles a variety of challenging problems well and has good tracking performance.

The present paper is organized as follows: In Section 2, studies on Siamese networks and correlation tracking are summarized, while in Section 3, a fully convolutional Siamese network is described for object tracking. Then, in Section 4, the performance comparison of experimental results between the proposed tracking algorithm and other latest tracking algorithms is presented. Lastly, in Section 5, the conclusion of this study and future research direction are presented.

2. Related Works

2.1. Correlation Filter-Based Tracking Algorithm

A correlation filter-based tracking method is a technique to train a discriminative classifier that can estimate an object's displacement between continuous frames. Learning samples are generated using the circular correlation characteristics around the target, and a correlation filter is trained by extracting shapes from the samples. This method has achieved a very effective improvement in various challenging tasks and benchmarks owing to its high computational efficiency and kernel trick method in the Fourier domain [17]. This method consists of a form of circular shifts of input signals to a target Gaussian function, which does not need hand-crafted features of the target. Generally, a correlation filter generates a correlation peak in each interested patch of the frame and produces a low response in the background region. This is used as a reference filter to identify a specific target. Using this filter, a tracking problem can be solved, but the filter has to be trained in real time. Due to this limitation, it is not suitable for online tracking; however, the minimum output sum of squared error (MOSSE) methodology has been researched to propose a new direction [12]. Studies on various algorithms, to which the adaptive learning method theory proposed by MOSSE was applied, have been conducted. The MOSSE filter was improved by exploiting the circulant structure with kernels [18]. The channel and spatial reliability concepts were applied to the discriminative correlation filter (DCF) tracking, with the spatial reliability map being used for the filter adjustment in the partial region of the target object [19]. The improved kernelized correlation filters employed multi-channel features, and are the most widely used filters based on their overall outstanding performance and high frame-per second rate [20]. A spatially regularized discriminative correlation filter (SRDCF) tracker imposes constraints on the correlation filter coefficients according to locations, using a spatial regularization component in training to induce boundary effects [13]. An MCCTH-Staple tracker combines various types of features and configures various experts through DCF for independently tracking the target object by each expert [21].

2.2. CNN-Based Tracking Algorithm

Deep learning has been used to obtain technical features as an emerging technology and has proven its excellent capability in various works in computer vision and pattern recognition such as image and video classification as well as object recognition [22,23]. For example, a CNN has been used in various computer vision problems such as image classification, semantic segmentation, and motion recognition due to its improved performance. More recently, studies on the application of CNN's advantages in tracking algorithms have been conducted. These tracking algorithms have combined deep feature maps with correlation filter trackers to improve tracking performance for better identification. DeepSRDCF [24] and FCNT [25] are used in object-tracking processes by extracting deep feature maps of many layers from the pre-trained model such as VGG or AlexNet. To ensure the accuracy and robustness of VOT, deep feature maps at different layers were used [26]. DeepTrack configures a number of CNN classifiers at instances of different objects to exclude noise samples during the model update, which is finely performed by adjusting a deep model online [27]. The key point that is noticeably shared by the above CNN-based trackers is as follows. First, the features produced in the last layer finely express the information represented in the object, and second, they are useful to accurately predict the object's location even if environmental changes inside the image occur. To use features at many CNN hierarchies to the highest extent, studies on designing a dual structure have been conducted to use hierarchical features at different layers in a deep model and obtain a better shape representation from various streams.

2.3. Siamese Network

A Siamese network shows excellent performance in the problem solving of face recognition and image matching, which is the similarity comparison area [28,29]. The structure of the Siamese network is shown in Figure 1.

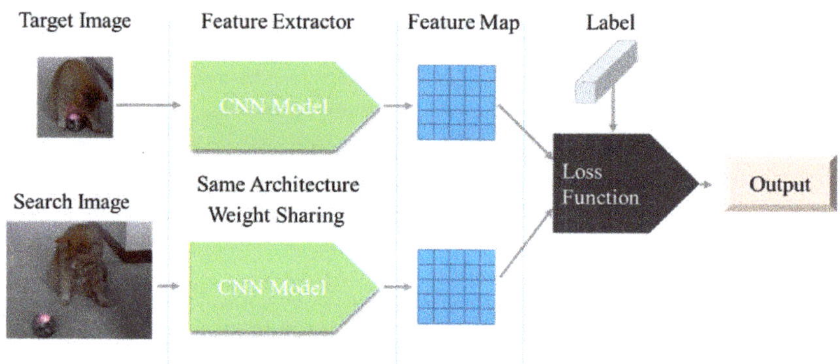

Figure 1. Standard Architecture of the Siamese Network.

The research on the application of the Siamese network to object-tracking problems, which is similar to the solution of the image similarity problem, has gained traction. Much attention has been paid to a Siamese network due to its tracking accuracy and speed balance performance. The tracking problem can be defined as a problem matching the appearance of the target object with the template image in the search region. For the input data used in a Siamese network, generally, the template image of the target object and images with or without the target object are used. The target object template is normally initialized at the first frame, with the same template used in continuous frames.

The pioneering works of the Siamese network tracker are SINT and SiamFC [30,31]. These two algorithms defined the tracking problem as the measurement of the target similarities between the first and current frames. SINT defined the tracking problem as the verification work to learn the similarity between inputs. These approaches have gained many important points due to the inherent performance of Siamese networks. However, if a network is trained with a small dataset, the overfitting problem may occur. A SiamFC used an embedded CNN model to extract input image features and fused them using a correlation layer to generate a response map. Follow-up studies have been conducted to improve the SiamFC, with CFNet [32] acting as an enhanced version of SiamFC, which is a closed-form solution. A correlation filter layer is applied within the template branch to improve the information that is contained in feature maps.

3. Proposed Method

In this section, the proposed network for tracking as shown in Figure 2 is described. The proposed network employs two images as the input specified as target object and searches images. The object region coordinates them, and information about the presence of the target object are extracted as the inputs pass through the fully CNN-based backbone network and RPN. The backbone network that extracts object features was designed with a customized structure and modified into a Siamese network form. In Figure 2b, weight sharing means that each kernel of the convolution layer has the same weight. Two images input to the network pass through the same network and output a value indicating similarity. At this time, if the weights are not shared, it is structurally the same network, but it is difficult to obtain the correct result for the input data because different weights are learned. Therefore, the network is learned using the loss value output in Figure 2c, and the weights have the same value in this process.

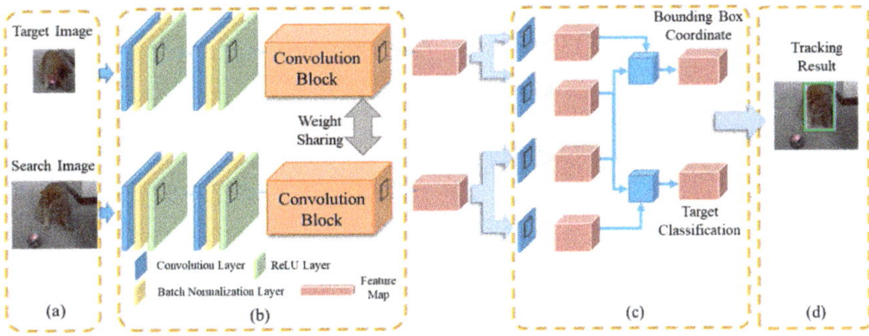

Figure 2. Proposed Tracking Algorithm with Siamese Network: (**a**) Target image and search region, (**b**) Siamese network consisting of convolution blocks used in feature extraction and highlighting the region of interest, (**c**) RPN for object region and bounding box coordinate prediction, (**d**) Tracking result marked by the green box.

3.1. Convolution Block for Feature Extraction

The most in-demand part of computation in a CNN is a fully connected layer, with the network proposed in this study shown as a full CNN, in which fully connected layers are removed and replaced with convolution layers. The computation amount in convolution layers increases with the number of kernels used for feature extraction. A convolution layer was designed by converting it to a bottleneck layer structure to reduce the computation amount. A bottleneck layer structure is effective in reducing the number of parameters by changing the internal structure of the network. In Equation (1), the number of parameters in the network is calculated. Figure 3 shows comparison of the No. of Parameters in a convolution layer.

$$\text{Parameters} = \text{In Channels} \times \text{Out Channels} \times \text{Kernel Width} \times \text{Kernel Height} \quad (1)$$

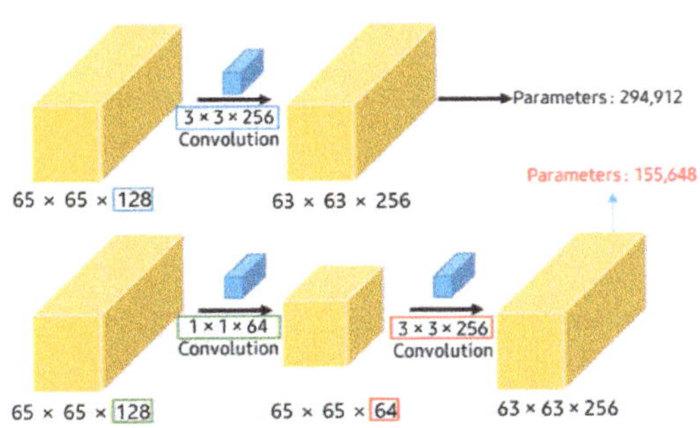

Figure 3. Comparison of the No. of Parameters in a Convolution Layer; Above: Standard convolution, Below: Bottleneck structured convolution.

The bottleneck structure consists of a three-step cycle of output compression, feature extraction using convolution, and output expansion. The output compression employs a 1 × 1 convolution. A kernel a size of 1 × 1 is used when adjusting the number of input feature maps. The output feature map in the previous layer is used as the input in the next layer. If feature maps are extracted using a smaller number of kernels than the number of input feature maps, the number of feature maps is reduced, thereby significantly decreasing the computation amount. In the feature extraction step, convolution is conducted using a kernel with a size of N × N. In the last output expansion step, the dimension is increased from a 1 × 1 to 3 × 3 convolution. Since the amount of computation and the number of parameters that are linked between layers are significantly reduced if a layer is designed with the bottleneck structure, a deeper network can be designed and learned with the same computing and time resource. Figure 4 shows the convolution block used in the proposed network, in which each block is composed of one or more bottleneck layers.

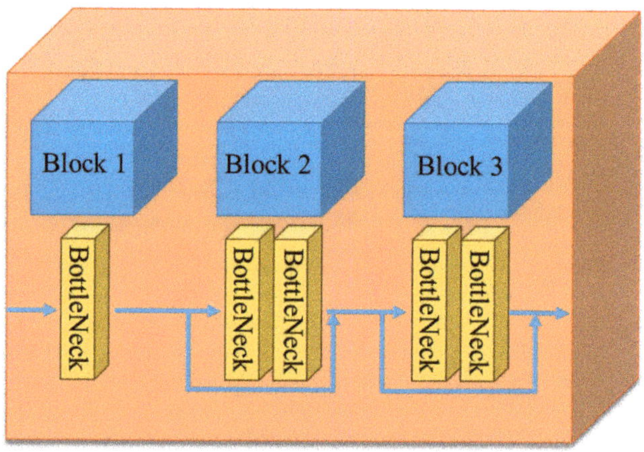

Figure 4. Structure of Convolution Block.

The bottlenecks were arranged reflecting the CNN's characteristic that extracts significant features in the bottlenecks near the input layer, followed by semantic features more and more near the end. To extract more semantic features, a convolution block structure consisting of bottlenecks was iterated to increase the number of kernels, aiming to extract feature information to the highest extent. If the continuous bottleneck structure is used, it is likely for there to be a loss of information. Thus, if more than two structures are iterated, the information flow is preserved by connecting the information in the input feature map to the output feature map.

3.2. *Siamese Network Architecture*

A Siamese network is a neural network architecture consisting of two or more of the same networks. A Siamese network shares the same parameters and weights. The parameter update can be performed by mirroring two sub-networks. Figure 5 shows the basic structure diagram of a Siamese network. By comparing feature vectors extracted from two input images, parameters are trained to find the similarity. In a general neural network, a method to predict many classes is trained. If a new class is added or removed to/from a dataset, a problem occurs. In such a case, the neural network has to be updated and the entire dataset should be re-trained. A large amount of data is needed to train a neural network, whereas a Siamese network can train a similarity function to verify whether the appearances of two images are the same.

Figure 5 shows the network structure used to solve the tracking problem. To increase the number of kernels that extract features, a convolution block is layered in the design. The final output feature map in the tracking object region is 18 × 18 × 256 in size, while the final output feature map in the search region is 34 × 34 × 256 in size.

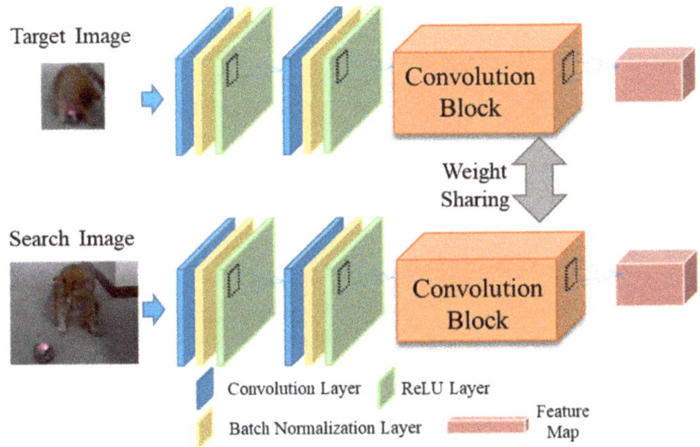

Figure 5. Proposed Siamese Network Architecture.

3.3. Region Proposal Network

The RPN is known as a very effective method for object detection. The main purpose of this network is to infer specific objects and regions that are present in the images. This network was introduced in the paper "Faster R-CNN: Towards Real-Time Object Detection with Region Proposal Networks" [33]. The region proposal is conducted by regressing the center coordinate of the anchor box in the regions where the object is likely to exist inside the image.

For the object region, a feature map that can be obtained in the last layer in the CNN is used, with Figure 6 showing the structure of the RPN. An anchor box is arranged in every cell of the feature map with a size of N × N. The number of anchor boxes used in the region proposal can be selected by a user, and the use of anchor boxes of various sizes has an advantage of inferring accurate regions. On the other hand, as the number of anchor boxes increases, so does the number of computations.

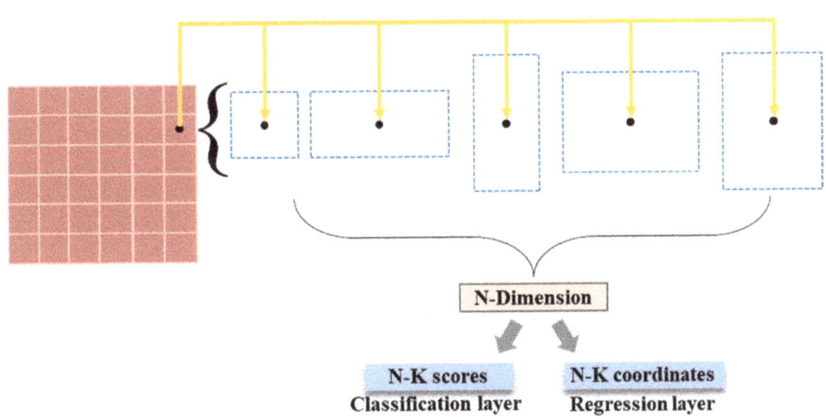

Figure 6. Anchor Box Shapes in the RPN.

The key of the RPN is to infer a coordinate of the anchor box through regression and determine whether an object is present. An anchor box includes four values of centerX, centerY, width, and height. The number of anchor boxes used in the inference is determined according to the box scale and aspect ratio. For example, if the scale is three, and the aspect ratio is three, nine anchor boxes are created. The created anchor boxes are positioned in each cell of the feature map. CenterX and centerY are fixed for each anchor box while the width and height are determined by a ratio of the bounding box width and height of the target object.

A probability of object existence is assigned to each created anchor box. If this is zero, no object is present, while a one means that the object is present. The number of probabilities is proportional to the number of anchor boxes. Let us assume that nine anchor boxes are assigned to a feature map of 17 × 17 in size. Then, the number of coordinates to be inferred is calculated as 17 × 17 × 9 × 4 = 10,404, and the number of anchor boxes to be created as 17 × 17 × 9 = 2601. Whether the object is present is determined using the final 2601 probabilities. The final region is assigned by combining anchor boxes where the object is present. In this study, one scale and five sizes of aspect ratios were designated, and the final number of anchor boxes was five.

Figure 7 shows the proposed RPN structure. For its input, a target object feature map of 18 × 18 × 256 in size, which was extracted through the Siamese network, and a search region feature map with a size of 34 × 34 × 256, were used. These feature maps were converted into four feature maps via the convolution layer, which extracts regression and object existence probability values.

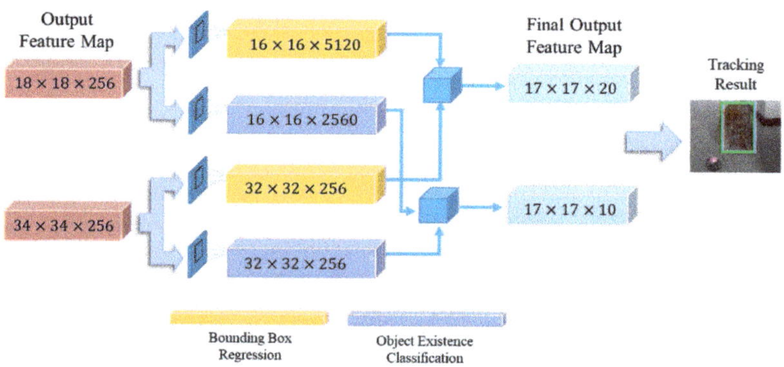

Figure 7. Proposed RPN Architecture.

At first, feature maps for the anchor box regression values for 256 feature maps were extracted for the target object. The number of obtained output feature maps was calculated as 256 × 5 × 4 = 5120. The values for determining whether the object is present for the same feature map were extracted. Then, 256 × 5 × 2 = 2560 values could be obtained. In the search region, the same numbers of input and output feature maps were applied and computed for the coordinate regression and object existence.

4. Experiments

4.1. Experimental Environment Configuration

The hardware specifications used in the experiment are as follows: the central processing unit used was the Intel Core i7 8-generation 8700 K series, and the graphic card was the NVIDIA TITAN X Pascal 12 GB series consisting of 3840 compute-unified device architecture (CUDA) cores. The read access memory and hard disk were a DDR4 48 GB and solid-state drive, respectively, so as to guarantee fast input and output. Table 1 indicates the detailed hardware specifications used in the experiment.

Table 1. Hardware Specifications.

	Detailed Specifications
CPU/RAM	Intel I7-8700K 3.7 GHz/DDR4 16 G
GPU	Geforce Titan Xp 12 GB 3840 CUDA cores
Storage	Samsung 512 G SSD

The operating system was Windows 10, along with the CUDA toolkit 11.1 version and the CUDA deep neural network library 8.0 version. To implement the tracking algorithm, Python 3.8 version was used and the deep learning framework used in the network design was PyTorch 1.8 version. Table 2 indicates the detailed software specifications used in the experiment.

Table 2. Software Specifications.

	Detailed Specifications
OS	Windows 10
Language	Python 3.8
GPU-accelerated libraries	CUDA 11.1/cuDNN 8.0
Deep Learning Framework	PyTorch 1.8

4.2. Dataset Configuration

In this study, the ILSVRC2017 VID dataset and object tracking benchmark (OTB) dataset were used. The datasets were divided into two parts, according to their purpose. The ILSVRC2017 VID dataset was used as the learning data to train the neural network of the proposed tracking algorithm, while the OTB dataset was employed to quantitatively evaluate the performance of the tracking algorithm [34,35].

4.2.1. ILSVRC 2017 VID Dataset

A variety of video-based datasets was proposed for various visual applications in the computer vision field. The ILSVRC 2017 VID dataset was the most represented benchmark dataset. This dataset is divided into train and validation sets consisting of 3862 video snippets and 555 video snippets, respectively. Figure 8a shows the train set in the ILSVRC VID dataset. Images extracted from each video snippet at a unit of one frame are used for training purposes. Figure 8b shows the frames extracted from each snippet video, with the number of frames being different from video to video.

(a) (b)

Figure 8. Information of ILSVRC 2017 VID Datasets: (**a**) Video snippets, (**b**) Extracted image frame of snippets.

The training and validation videos are extracted at a unit of frame. Each frame was matched with a 1:1 annotation, and videos can be controlled through this annotation. The key elements in the annotation are size and bndbox. The size refers to the size of the frame. Using this information, a conversion is conducted into a certain ratio size suitable for learning. "bndbox" has four attributes and refers to the location information of the object present in the frame. "xmin" and "ymin" refer to the upper left corner of the rectangular bounding box, while "xmax" and "ymax" refer to the lower right corner.

4.2.2. Object Tracking Benchmark Dataset

The OTB dataset consists of three datasets, namely OTB-2013, OTB-50, and OTB-100. OTB-2013 is composed of 50 video sequences for the quantitative evaluation of VOT algorithms. OTB-100 contains an additional 50 videos in addition to the OTB-2013 dataset, resulting in 100 videos. OTB-50 is produced by selectively extracting 50 videos whose tracking is relatively difficult out of the OTB-100 dataset.

The video sequences included in the OTB benchmark dataset include 11 types of different attributes such as illumination variation (IV), scale variation (SV), and occlusion (OCC). A video may have multiple attributes instead of a single one. Table 3 presents the detailed description of each attribute, and Figure 9 shows the video distribution graph about attributes. The count in the lower end indicates the number of videos assigned to each attribute.

Table 3. Sequence Attribute Table for Algorithm Evaluations.

Attribute	Description
Illumination variation (IV)	Illumination variation in the target object region
Scale variation (SV)	Scale variation in the tracking object
Occlusion (OCC)	Occlusion generated in the target object region
Deformation (DEF)	Non-rigid deformation of the object
Motion blur (MB)	Motion blur occurred in the target object
Fast motion (FM)	Fast motion of the object detected
In-plane rotation (IPR)	Object rotation detected in the image
Out-of-plane rotation (OPR)	Object rotation detected outside the image
Out-of-view (OV)	A region of the object moved outside the image
Background clutters (BC)	Color or texture created similar to the object
Low resolution (LR)	The low resolution of the object

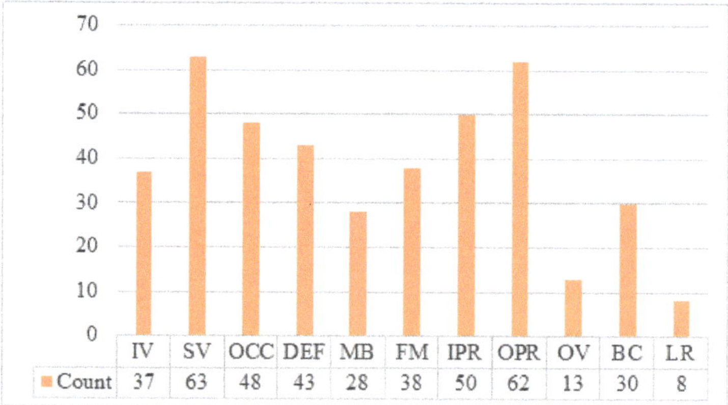

Figure 9. Sequence Distribution Graph about Attributes.

Figure 10 shows the JSON file format in the OTB dataset used in the evaluation. "video_dir" refers to the dataset folder name; "init_rect" refers to the initial region coordinate; "img_names" refers to the image file name in the folder; "gt_rect" refers to the coordinate value designated in the manual; and "attr" refers to the attribute of the video.

```
{
  "Basketball": {
    "video_dir": "Basketball",
    "init_rect": [198, 214, 34, 81],
    "img_names": [↔],
    "gt_rect": [↔],
    "attr": ["Illumination Variation", "Out-of-Plane Rotation", "Occlusion", "Deformation", "Background Clutters"]
  },
```


Figure 10. Annotation Format of OTB Dataset.

4.3. Network Training and Testing

4.3.1. Data Preprocessing and Labeling

The ILSVRC 2017 training dataset was used to train the proposed network model. This training dataset consisted of 3862 videos and 1,122,397 images that were extracted from the videos at a unit of frame. Each frame was defined by time-series image data, and the network employed four label parameters of target image, search region, normalized bounding box coordinate, and object existence. Each data should be processed to be used as the input in the network. Two videos that were randomly chosen from each video frame in the training dataset were extracted as a pair. Since training was conducted through a comparison of similarities between objects in the network, the chronological order of the images was ignored in the data load process. Additionally, each image was used as a search region and a target image after preprocessing.

Figure 11 shows a pair of preprocessed training images. The target object is located on the left side of the images in Figure 11a,b. The preprocessing procedure of the image is as follows: for the target image, a margin of 15 to 25% in the object region area is randomly added to expand the image. The image boundary is set to the maximum margin. In the search region, the area of the object region and height are randomly reduced or expanded within a ratio of 90 to 110% of the original size.

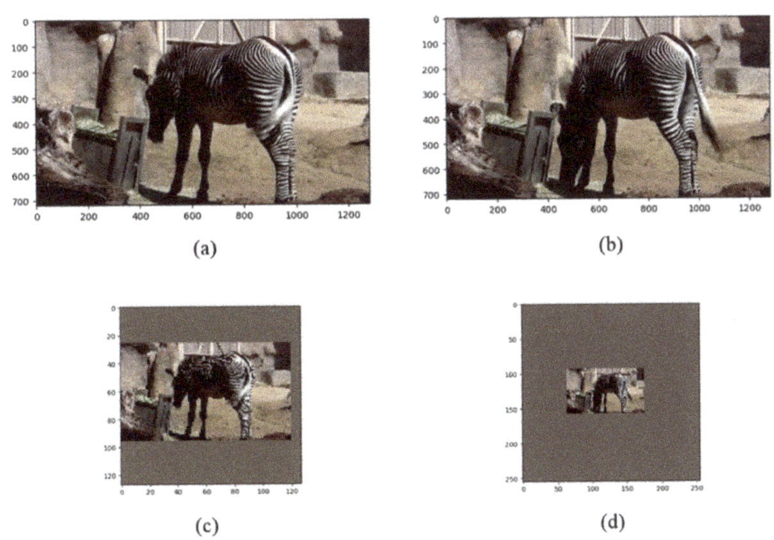

Figure 11. Preprocessed Training Dataset. (**a**) Original target image, (**b**) Original search region, (**c**) Converted target image, (**d**) Converted search image.

Figure 11c,d show the final images after preprocessing. They are reconfigured so that the center point in the object region is positioned at the center of the image. The target image and search region were converted to 127 × 127 and 255 × 255 in size, respectively, for network inputs. In the size conversion, the margin was cut while maintaining the image ratio to preserve the shape of the object.

In the preprocessing procedure, the object coordinates in the original image and reconfigured image were changed. Reprocessing was conducted by reflecting the preprocessed image coordinate information in the normalized bounding box coordinate and object existence labels.

The four-dimensional score map was the final output of the network (4D) (N, 20, 17, 17). The first dimension refers to the mini arrangement size, the second dimension to the anchor box coordinate, and the third and fourth dimensions to the size of the score map.

In the labeling work, 2D, 3D, and 4D data were used. The 2D anchor box coordinate data consisted of 20 records. Each of the four coordinates of x, y, width, and height had five anchor boxes. The 3D and 4D score map size was 17×17 in size with 1445 anchor boxes in total. The anchor box label for regression was also made with the same size as that of the score map. The coordinate of each anchor box can be acquired by calculating the target image coordinate in the scale-adjusted search region and anchor box coordinate.

$$Regression_x = \frac{GT_x - anchor_x}{anchor_w} \tag{2}$$

$$Regression_y = \frac{GT_y - anchor_y}{anchor_h} \tag{3}$$

$$Regression_w = \ln\left(\frac{GT_w}{anchor_w}\right) \tag{4}$$

$$Regression_h = \ln\left(\frac{GT_h}{anchor_h}\right) \tag{5}$$

Equations (2) to (5) show the normalized anchor box coordinates to be estimated in the network [36]. Equations (2) and (3) are formulas for normalizing the center point coordinates of the anchor box, Equations (4) and (5) are formulas for normalizing the width and height of the anchor box. The normalized coordinates are used as labels, and the network is trained by applying the smooth L1 loss function to this value.

The classification label was used to determine whether the target object existed inside the anchor box. If the object did exist, a one, otherwise zero, and other than that, -1 was assigned. The object's existence was determined according to the intersection ratio result between each anchor box and ground truth (GT) region.

The intersection over union (IOU) was used to calculate the overlap rate. If the IOU was more than 50%, it would determine that the object did exist, thereby assigning a 1 to the anchor box. If the IOU was less than 40%, it would determine that the object did not exist in the anchor box, thereby assigning a 0 to the anchor box. If the IOU was between 40% and 50%, it determined that the object's existence was unclear. In that case, a -1 was assigned to not affect the weight training. The created number of classification labels was 1445, which is the same as the number of anchor boxes. Each label was used to train the weight in the network. To train the proposed network, the following values were used as hyper-parameters. The optimizer used Adam and set learning rate = 0.001, coefficient for primary momentum = 0.9, coefficient for secondary momentum = 0.999, and epsilon = 10^{-8}, weight_decay = 0.01.

4.3.2. Loss Function for Network Training

Two types of loss functions were used in the calculation for the network model, which was the key to the tracking algorithm. One was the classification loss to determine whether a tracking object was present inside the anchor box and the other was the regression loss of the anchor box coordinates to estimate the object region. The classification loss only classifies whether objects are included without specific classification of what class the anchor box is. If the object exists, 1 is assigned as positive, and 0 is assigned as negative if it does not exist. The regression loss predicts the coordinates of the anchor box. The coordinates are x and y, which are the center coordinates of the box, as well as width and height, which are the box size. The proposed network uses five anchor boxes. The classification labels obtain 2 (positive, negative) × 5 (anchor box) = 10 values, and the regression labels obtain 4 (x,y,w,h) × 5 (anchor box) = 20 values. Each value is assigned to each grid of the last feature map. Each of the loss functions was calculated using the prediction value and label in the network. To determine the object's existence, it was assumed as a classification problem to use the cross-entropy function. Equation (6) presents the cross-entropy function.

$$\begin{aligned} loss_{cls} &= -log\left(\frac{\exp(x[class])}{\sum_j \exp(x[j])}\right) \\ &= -x[class] + log\left(\sum_j \exp(x[j])\right) \end{aligned} \tag{6}$$

In Equation (6), the anchor box is classified whether or not objects are included without specific classification. If the object exists, 1 is assigned as positive, and 0 is assigned as negative if it does not exist.

SmoothL1Loss was used for the regression loss of the anchor box coordinate. Equation (7) presents the SmoothL1Loss equation. In this equation, β refers to the hyperparameter, which is generally defined as one.

$$Smooth_{L_1-reg} = \begin{cases} \frac{0.5(x_n - y_n)^2}{\beta}, & if\ |x_n - y_n| < \beta \\ |x_n - y_n| - 0.5 \times \beta, & otherwise \end{cases} \tag{7}$$

In Equation (7), $|x_n - y_n|$ value is smaller than β term, so a square term is used. Otherwise, the following L_{1-reg} term is used. Due to this characteristic, it is less sensitive to abnormal values and prevents gradient exploding. The final loss function is calculated by summing Equations (6) and (7), which are expressed in Equation (8).

$$loss_{tot} = loss_{cls} + Smooth_{L_1-reg} \tag{8}$$

4.4. Quantitative Evaluation Index

OTB-50 and OTB-100 tracking benchmark datasets were used for performance evaluations. OTB-50 consisted of 50 videos and 29,500 frames. OTB-100 had an additional 50 videos compared to OTB-50, and included 100 videos and 50,000 frames. The videos included in the benchmark had various attributes. For the quantitative evaluation of the proposed algorithm, the one-pass evaluation (OPE) index was used.

In the OPE, two types of evaluation indices were present, namely precision and success plots. In the precision plot, the center position error was calculated. The center position was used by the reference of the GT coordinate provided in the manual. The coordinate center position produced by the tracking algorithm was calculated, and the calculated value was compared with the GT's center position.

The success plot refers to an overlap rate of bounding boxes that surrounded the object region. To calculate an overlap rate, manual-coordinate GT and object-coordinate system truth (ST) extracted through the tracking algorithm were used. The coordinate format consisted of four values indicated by x and y coordinates, which indicated the left top, width, and height. These coordinate values were used to calculate the overlap rate of the bounding boxes.

As shown in Figure 12, GT_Box refers to a bounding box region consisting of GT coordinates, and ST_Box refers to a bounding box region of the object produced by tracking with the user's tracking algorithm. The overlap rate is calculated by Equation (9) below.

$$IoU(\text{GT\_Box}, \text{ST\_Box}) = \frac{Area(\text{GT\_Box} \cap \text{ST\_Box})}{Area(\text{GT\_Box} \cup \text{ST\_Box})} \tag{9}$$

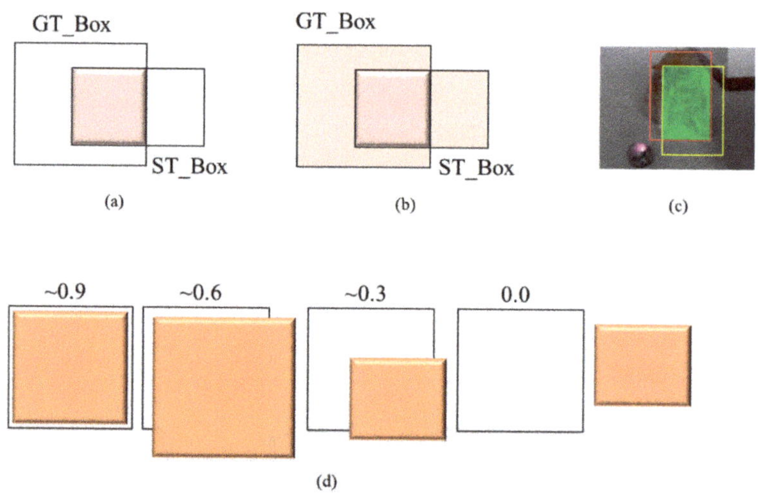

Figure 12. Calculation Method of Intersection Ratio Using Bounding Boxes. (**a**) Intersection, (**b**) Union, (**c**) Applied Image, (**d**) Numerical Change of Overlap Rate in the Overlap Region.

The denominator in Equation (9) is a union region of GT_Box and ST_Box in Figure 12b. The numerator is an intersection region of GT_Box and ST_Box in Figure 12a. The result of IOU is the region filled with the green-color box in Figure 12c. To express the performance rank, the area under a curve was used.

4.5. Experimental Results

In this study, the performance of the tracking algorithm was evaluated using two quantitative evaluation indicators, success plot and precision plot. The success plot measures the overlap ratio of the bounding box suggested by the algorithm and the GT box area measured by the manual method. The precision plot measures the distance difference between the center point of the bounding box presented by the algorithm and the center point of the GT box measured by the manual method. This metric indicates the accuracy with which the tracking algorithm continuously tracks the object. To quantitatively compare the performance of the proposed tracking algorithm, four tracking algorithms, namely BACF [9], CSRDCF-LP [19], DCF [20], and MCCTH-Staple [21] were used. The colors in the produced graph were red, green, magenta, yellow, and sky blue from the first to fifth ranks, respectively. The line shapes in the graph comprised a line, dash, dot, line, and dash, in that order. To ensure the statistical validity of the tracking results, the proposed algorithm and comparison algorithm were tested using the same experimental environment described in Section 4.1.

Figure 13 and Table 4 present the performance evaluation results using the OTB-50 benchmark dataset, with the proposed algorithm achieving the highest scores of 0.572 and 0.799.

Table 4. Overall Performance Evaluation Results of OTB-50.

	Proposed	**BACF**	**MCCTH-Staple**	**DCF**	**CSRDCF-LP**
Success	**0.572**	0.55	0.549	0.408	0.502
Precision	**0.799**	0.757	0.713	0.606	0.662

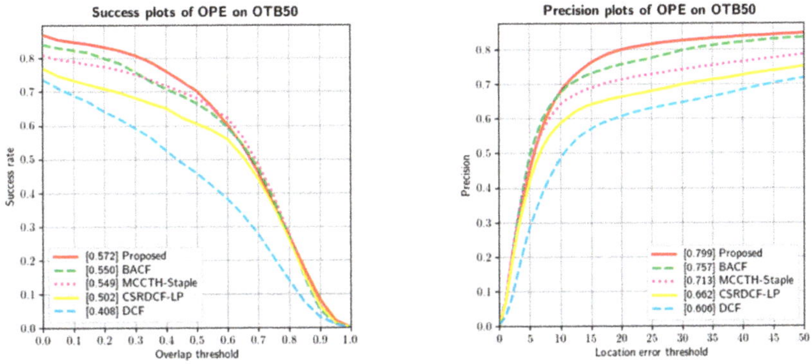

(a) (b)

Figure 13. Evaluation Graph of OTB-50 for All Attributes (**a**) Success Plots Graph of OTB-50, (**b**) Precision Plots Graph of OTB-50.

Figure 14 and Table 5 present the performance evaluation results using the OTB-100 benchmark dataset. The proposed algorithm exhibited the highest scores as it achieved 0.633 and 0.847. Tables 6 and 7 summarize all the results using the OTB dataset. The bold font was used for the highest values. Table 6 presents the results using the OTB-50 dataset. The success plot result, which showed an overlap rate with the OTB-50 dataset, achieved 0.572. This result was higher than that of the DCF algorithm, which showed the lowest result, by 0.164. It showed a slight difference (0.022) compared to BACF, which achieved the second-highest result. The precision plot achieved 0.799, which showed the center error. This result was higher than that of the DCF algorithm, which showed the lowest result, by 0.193. The BACF algorithm, which achieved the second-highest value as the same as shown in the success plot result, exhibited 0.757.

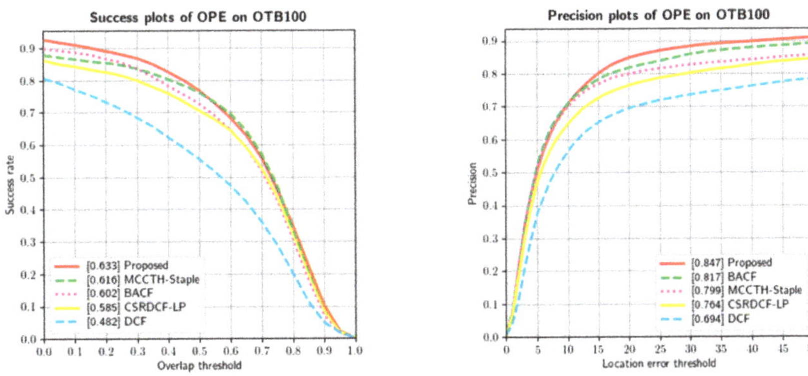

(a) (b)

Figure 14. Evaluation Graph of OTB-100 for All Attributes (**a**) Success Plots Graph of OTB-100, (**b**) Precision Plots Graph of OTB-100.

Table 5. Overall performance evaluation results of OTB-100.

	Proposed	BACF	MCCTH-Staple	DCF	CSRDCF-LP
Success	**0.633**	0.602	0.616	0.482	0.585
Precision	**0.847**	0.817	0.799	0.694	0.764

Table 6. Overall Performance Evaluation Results for Attributes of OTB-50 Benchmark Dataset.

		Proposed	BACF	MCCTH-Staple	DCF	CSRDCF-LP
Total	Success	**0.572**	0.550	0.549	0.408	0.502
	Precision	**0.799**	0.757	0.713	0.606	0.662
IPR	Success	0.518	**0.540**	0.514	0.398	0.462
	Precision	0.741	**0.748**	0.683	0.572	0.607
OCC	Success	**0.564**	0.516	0.552	0.389	0.464
	Precision	**0.801**	0.708	0.715	0.581	0.608
OV	Success	**0.548**	0.483	0.491	0.328	0.435
	Precision	**0.760**	0.704	0.671	0.443	0.624
IV	Success	0.557	**0.587**	0.549	0.451	0.466
	Precision	0.766	**0.792**	0.724	0.688	0.607
LR	Success	0.520	0.437	**0.571**	0.255	0.486
	Precision	0.821	0.695	**0.834**	0.543	0.711
BC	Success	0.578	**0.585**	0.517	0.437	0.445
	Precision	0.772	**0.797**	0.679	0.640	0.575
FM	Success	**0.569**	0.534	0.524	0.407	0.536
	Precision	0.746	**0.749**	0.646	0.567	0.693
MB	Success	**0.570**	0.542	0.492	0.422	0.535
	Precision	**0.767**	0.756	0.625	0.589	0.692
SV	Success	**0.563**	0.506	0.525	0.366	0.470
	Precision	**0.786**	0.710	0.680	0.569	0.622
DEF	Success	**0.536**	0.514	0.530	0.413	0.493
	Precision	**0.751**	0.710	0.692	0.612	0.688
OPR	Success	**0.541**	0.518	0.536	0.394	0.432
	Precision	**0.762**	0.719	0.694	0.573	0.562

Table 7. Overall Performance Evaluation Results for Attributes of OTB-100 Benchmark Dataset.

		Proposed	BACF	MCCTH-Staple	DCF	CSRDCF-LP
Total	Success	**0.633**	0.602	0.616	0.482	0.585
	Precision	**0.847**	0.817	0.799	0.694	0.764
IPR	Success	0.558	**0.567**	0.565	0.471	0.534
	Precision	0.783	**0.792**	0.754	0.678	0.716
OCC	Success	**0.616**	0.560	0.595	0.441	0.527
	Precision	**0.825**	0.756	0.756	0.619	0.671
OV	Success	**0.595**	0.483	0.491	0.328	0.435
	Precision	**0.796**	0.704	0.671	0.443	0.624
IV	Success	**0.642**	0.627	0.592	0.484	0.564
	Precision	0.828	**0.830**	0.755	0.727	0.721
LR	Success	0.501	0.446	**0.572**	0.264	0.495
	Precision	0.829	0.729	**0.851**	0.584	0.743

Table 7. *Cont.*

		Proposed	BACF	MCCTH-Staple	DCF	CSRDCF-LP
BC	Success	0.636	**0.646**	0.610	0.513	0.553
	Precision	0.843	**0.863**	0.788	0.724	0.715
FM	Success	**0.614**	0.572	0.580	0.464	0.593
	Precision	**0.790**	0.782	0.713	0.632	0.759
MB	Success	**0.628**	0.584	0.570	0.470	0.584
	Precision	**0.807**	0.777	0.705	0.614	0.732
SV	Success	**0.598**	0.538	0.585	0.397	0.537
	Precision	**0.808**	0.766	0.766	0.623	0.710
DEF	Success	**0.596**	0.555	0.586	0.450	0.544
	Precision	**0.811**	0.777	0.779	0.645	0.747
OPR	Success	**0.596**	0.518	0.536	0.394	0.432
	Precision	**0.816**	0.719	0.694	0.573	0.562

Table 7 presents the results using the OTB-100 dataset, which achieved a 0.621 in the success plot as the highest result. The algorithm that exhibited the lowest result was also the DCF algorithm, which was the same as the result of the OTB-50 dataset. The algorithm that exhibited the second-highest result with a 0.633 achieved was MCCTH-Staple, which was different from that using the OTB-50 dataset. BACF, which achieved the second-highest value using the OTB-50, showed the third-highest result with a 0.602.

The proposed algorithm achieved the highest result with a 0.847 in the precision plot. In contrast to the success plot, the BACF algorithm achieved the second-highest result with a 0.817.

The MB attribute showed high success plot results out of 11 attributes in the OTB-50 and OTB-100 datasets. Figure 15 shows the first frame in the Tiger1 video with the MB attribute, and this video contains the IV and FM attributes as well. Overall, these three attributes showed good results in the OTB-50 and OTB-100 datasets.

Figure 15. MB Attribute—First Frame of Tiger 1 Sequence.

Figure 16 depicts the frame that includes all MB, IV, and FM attributes. Although afterimages caused by sudden object changes and blurring of pixels were found, it showed a robust tracking result even in the MB, IV, and FM attributes.

Figure 16. Frame Comprising the MB, IV, and FM Attributes.

Figure 17 shows five sequences, including attributes other than MB, IV, and FM. Figure 17a–e show the sequences of Box, BlurCar4, BlurOwl, Bird2, and Coke. As shown in Figure 17, sequences a, d, and e include an occlusion attribute that hides the target object. The proposed tracking algorithm (red bounding box) exhibited good tracking without any drift that missed the object.

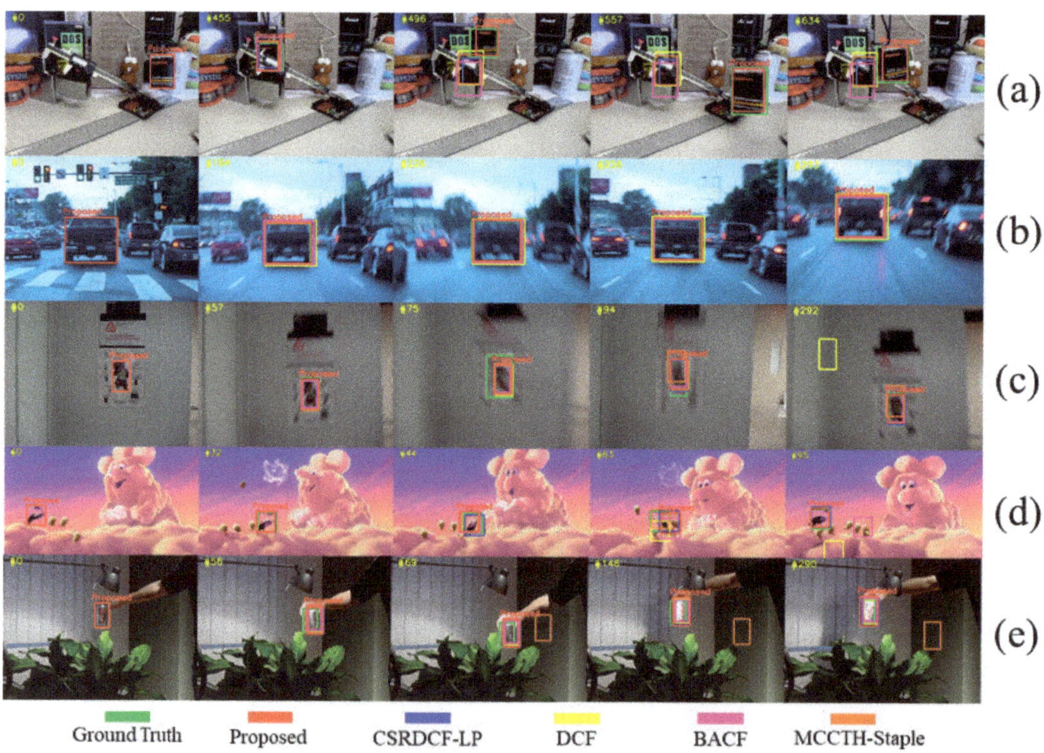

Figure 17. Tracking Results of Various Tracking Sequences. (**a**) Box Tracking Result, (**b**) BlurCar4 Tracking Result, (**c**) BlurOwl Tracking Result, (**d**) Bird2 Tracking Result, (**e**) Coke Tracking Result.

However, in Figure 17a, CSRDCF-LP, DCF, BACF, and MCCTH-Staple algorithms failed to track at 496 frames. In frame 455, it can be seen that the target object area is mostly obscured by obstacles. In the process of updating the object-appearance model based on the object area of this frame, it is judged that tracking failed because the appearance model was contaminated by noise (e.g., obstacle image information, background information) rather than by the target object. A similar situation can be seen in Figure 17d. In frame 63, it can be seen that all tracking algorithms localize an area other than the green GT area. In frame 95, the DCF and BACF algorithms showed the result of tracking failure, but it can be seen that the proposed algorithm tracks normally. When contamination by noise occurs in the external model, the value is continuously accumulated, so it is difficult to guarantee reliability in the comparison process. On the other hand, since the proposed algorithm is tracked through similarity comparison, re-tracking is possible when a place with high similarity between regions reappears. Figure 17b,c are images with blur caused by camera shake. In the case of Figure 17b, since the target object area is wider than in Figure 17c, it can be seen that all algorithms track relatively robustly even when blurring occurs. However, Figure 17c is more affected by the damage of the surrounding background caused by fast motion because the target object area is small. As a result, the DCF algorithm, which is highly affected by noise, failed to track. On the other hand, it shows that the proposed algorithm tracks the object well even if the object and the surrounding background are corrupted due to blurring caused by camera shake.

5. Conclusions

This study proposed a real-time single object tracking algorithm based on a Siamese network from two viewpoints of object tracking overlap accuracy and center tracking error rate. The tracking algorithm was a full CNN structure where the fully connected layers were removed to maintain spatial feature information, which was designed as a customized network. The Siamese network for feature extraction obtained the production of various feature maps using kernels 1×1 and 3×3 in size in the convolution layer. It was devised not to stop the feature flow by preventing a feature loss while the feature map produced in the bottleneck structure and input feature maps were connected and merged. The RPN estimated the object's location through the supervised learning method and determined whether the object was present in the bounding box. Through this mechanism, the potential location of the target object could be identified in the entire image frame. The experiment showed that the proposed algorithm achieved competitive results in some video attributes compared to other tracking algorithms. However, for low resolution and scale variation attributes, the proposed algorithm still had its limitations. To solve this object-tracking problem, future research is needed to overcome this drawback. To do this, studies on performance improvements are on track to be conducted by implementing feature representation techniques as additional functions that can finely extract detailed regions rather than large appearance areas.

Author Contributions: Conceptualization, S.-C.L., J.-H.H. and J.-C.K.; Data curation, S.-C.L. and J.-C.K.; Formal analysis, S.-C.L. and J.-H.H.; Investigation, S.-C.L., J.-H.H. and J.-C.K.; Methodology, S.-C.L. and J.-C.K.; Project administration, S.-C.L., J.-H.H. and J.-C.K.; Resources, S.-C.L., J.-H.H. and J.-C.K.; Software, S.-C.L., J.-H.H. and J.-C.K.; Supervision, S.-C.L. and J.-C.K.; Validation, S.-C.L. and J.-C.K.; Visualization, S.-C.L., J.-H.H. and J.-C.K.; Writing—original draft, S.-C.L., J.-H.H. and J.-C.K.; Writing—review and editing, S.-C.L., J.-H.H. and J.-C.K. All authors have read and agreed to the published version of the manuscript.

Funding: This research was supported by Basic Science Research Program through the National Research Foundation of Korea (NRF) funded by the Ministry of Education (No. 2021R1I1A1A01044223).

Institutional Review Board Statement: Not applicable.

Informed Consent Statement: Not applicable.

Data Availability Statement: Not applicable.

Conflicts of Interest: The authors declare no conflict of interest.

References

1. Kim, S.H.; Lim, S.C.; Kim, D.Y. Intelligent Intrusion Detection System Featuring a Virtual Fence, Active Intruder Detection, Classification, Tracking. *Ann. Nucl. Energy* **2018**, *112*, 845–855. [CrossRef]
2. Koubâa, A.; Qureshi, B. DroneTrack: Cloud-Based Real-Time Object Tracking Using Unmanned Aerial Vehicles over the Internet. *IEEE Access* **2018**, *6*, 13810–13824. [CrossRef]
3. Lee, S.J.; Lee, M.C. A Vision Based People Tracking and Following for Mobile Robots Using CAMSHIFT and KLT Feature Tracker. *J. Korea Multimed. Soc.* **2014**, *17*, 787–796. [CrossRef]
4. Laurense, V.A.; Goh, J.Y.; Gerdes, J.C. Path-Tracking for Autonomous Vehicles at the Limit of Friction. In Proceedings of the American Control Conference IEEE, Seattle, WA, USA, 24–26 May 2017; pp. 5586–5591.
5. Ma, C.; Huang, J.B.; Yang, X.; Yang, M.H. Hierarchical convolutional features for visual tracking. In Proceedings of the IEEE International Conference on Computer Vision, ICCV 2015, Santiago, Chile, 7–13 December 2015; pp. 3074–3082.
6. Li, B.; Wu, W.; Wang, Q.; Zhang, F.; Xing, J.; Yan, J. Siamrpn++: Evolution of siamese visual tracking with very deep networks. In Proceedings of the IEEE Conference on Computer Vision and Pattern Recognition, CVPR 2019, Long Beach, CA, USA, 15–20 June 2019; pp. 4282–4291.
7. Wang, N.; Shi, J.; Yeung, D.Y.; Jia, J. Understanding and diagnosing visual tracking systems. In Proceedings of the IEEE International Conference on Computer Vision 2019, Seoul, Korea, 27–28 October 2019; pp. 3101–3109.
8. Du, X.; Clancy, N.; Arya, S.; Hanna, G.B.; Kelly, J.; Elson, D.S.; Stoyanov, D. Robust Surface Tracking Combining Features, Intensity and Illumination Compensation. *Int. J. Comput. Assist. Radiol. Surg.* **2015**, *10*, 1915–1926. [CrossRef] [PubMed]
9. Kiani Galoogahi, H.; Fagg, A.; Lucey, S. Learning Background-Aware Correlation Filters for Visual Tracking. In Proceedings of the IEEE International Conference on Computer Vision, ICCV 2017, Venice, Italy, 22–29 October 2017; pp. 1135–1143.
10. Song, Z.; Sun, J.; Yu, J. Object Tracking by a Combination of Discriminative Global and Generative Multi-Scale Local Models. *Information* **2017**, *8*, 43. [CrossRef]
11. Li, P.; Wang, D.; Wang, L.; Lu, H. Deep visual tracking: Review and experimental comparison. *Pattern Recognit.* **2018**, *76*, 323–338. [CrossRef]
12. Bolme, D.S.; Beveridge, J.R.; Draper, B.A.; Lui, Y.M. Visual object tracking using adaptive correlation filters. In Proceedings of the 2010 IEEE Computer Society Conference on Computer Vision and Pattern Recognition 2010, San Francisco, CA, USA, 13–18 June 2010; pp. 2544–2550.
13. Danelljan, M.; Hager, G.; Shahbaz Khan, F.; Felsberg, M. Learning spatially regularized correlation filters for visual tracking. In Proceedings of the IEEE International Conference on Computer Vision 2015, Santiago, Chile, 7–13 December 2015; pp. 4310–4318.
14. Kiani Galoogahi, H.; Sim, T.; Lucey, S. Correlation filters with limited boundaries. In Proceedings of the IEEE Conference on Computer Vision and Pattern Recognition 2015, Boston, MA, USA, 7–12 June 2015; pp. 4630–4638.
15. Danelljan, M.; Robinson, A.; Khan, F.S.; Felsberg, M. Beyond correlation filters: Learning continuous convolution operators for visual tracking. In Proceedings of the European Conference on Computer Vision 2016, Munich, Germany, 8–14 September 2016; pp. 472–488.
16. Danelljan, M.; Bhat, G.; Shahbaz Khan, F.; Felsberg, M. Eco: Efficient convolution operators for tracking. In Proceedings of the IEEE Conference on Computer Vision and Pattern Recognition 2017, Honolulu, HI, USA, 21–26 July 2017; pp. 6638–6646.
17. Hadfield, S.J.; Lebeda, K.; Bowden, R. The visual object tracking VOT2014 challenge results. In Proceedings of the European Conference on Computer Vision (ECCV) Visual Object Tracking Challenge Workshop 2014, Zurich, Switzerland, 6–7 and 12 September 2014.
18. Henriques, J.F.; Caseiro, R.; Martins, P.; Batista, J. Exploiting the Circulant Structure of Tracking-by-Detection with Kernels. In Proceedings of the European Conference on Computer Vision, Florence, Italy, 7–13 October 2012; Volume 7575, pp. 702–715.
19. Lukezic, A.; Vojir, T.; Cehovin Zajc, L.; Matas, J.; Kristan, M. Discriminative correlation filter with channel and spatial reliability. In Proceedings of the IEEE Conference on Computer Vision and Pattern Recognition, CVPR 2017, Honolulu, HI, USA, 21–26 July 2017; pp. 6309–6318.
20. Henriques, J.F.; Caseiro, R.; Martins, P.; Batista, J. High-speed tracking with kernelized correlation filters. *IEEE Trans. Pattern Anal. Mach. Intell.* **2015**, *37*, 583–596. [CrossRef] [PubMed]
21. Wang, N.; Zhou, W.; Tian, Q.; Hong, R.; Wang, M.; Li, H. Multi-cue correlation filters for robust visual tracking. In Proceedings of the IEEE Conference on Computer Vision and Pattern Recognition 2018, Salt Lake City, UT, USA, 18–23 June 2018; pp. 4844–4853.
22. Liu, T.; Tao, D.; Song, M.; Maybank, S.J. Algorithm-dependent generalization bounds for multi-task learning. In Proceedings of the IEEE Transactions on Pattern Analysis and Machine Intelligence 2016, Washington, WA, USA, 3 March 2016; IEEE Computer Society: Pittsburgh, PA, USA, 2016; Volume 39, pp. 227–241.
23. Kim, J.-C.; Lim, S.-C.; Choi, J.; Huh, J.-H. Review for Examining the Oxidation Process of the Moon Using Generative Adversarial Networks: Focusing on Landscape of Moon. *Electronics* **2022**, *11*, 1303. [CrossRef]
24. Danelljan, M.; Hager, G.; Shahbaz Khan, F.; Felsberg, M. Convolutional features for correlation filter based visual tracking. In Proceedings of the IEEE International Conference on Computer Vision Workshops, ICCV 2015, Santiago, Chile, 7–13 December 2015; pp. 58–66.

25. Wang, L.; Ouyang, W.; Wang, X.; Lu, H. Visual tracking with fully convolutional networks. In Proceedings of the IEEE International Conference on Computer Vision, ICCV 2015, Santiago, Chile, 7–13 December 2015; pp. 3119–3127.
26. Qi, Y.; Zhang, S.; Qin, L.; Yao, H.; Huang, Q.; Lim, J.; Yang, M.H. Hedged deep tracking. In Proceedings of the IEEE Conference on Computer Vision and Pattern Recognition 2016, Las Vegas, NV, USA, 27–30 June 2016; pp. 4303–4311.
27. Li, H.; Li, Y.; Porikli, F. Deeptrack: Learning discriminative feature representations by convolutional neural networks for visual tracking. In Proceedings of the British Machine Vision Conference 2014, Nottingham, UK, 1–5 September 2014; Volume 1, p. 3.
28. Brandao, P.; Mazomenos, E.; Stoyanov, D. Widening siamese architectures for stereo matching. *Pattern Recognit. Lett.* **2019**, *120*, 75–81. [CrossRef] [PubMed]
29. Lin, T.Y.; Cui, Y.; Belongie, S.; Hays, J.; Tech, C. Learning Deep Representations for Ground-to-Aerial Geolocalization. In Proceedings of the IEEE Conference on Computer Vision and Pattern Recognition 2015, Boston, MA, USA, 7–12 June 2015; pp. 5007–5015.
30. Tao, R.; Gavves, E.; Smeulders, A.W. Siamese Instance Search for Tracking. In Proceedings of the IEEE Conference on Computer Vision and Pattern Recognition 2016, Las Vegas, NV, USA, 27–30 June 2016; pp. 1420–1429.
31. Bertinetto, L.; Valmadre, J.; Henriques, J.F.; Vedaldi, A.; Torr, P.H. Fully-Convolutional Siamese Networks for Object Tracking. In Proceedings of the European Conference on Computer Vision 2016, Munich, Germany, 8–14 September 2016; pp. 850–865.
32. Xu, H.; Gao, Y.; Yu, F.; Darrell, T. End-To-End Learning of Driving Models from Large-Scale Video Datasets. In Proceedings of the IEEE Conference on Computer Vision and Pattern Recognition, ICCV 2017, Venice, Italy, 22–29 October 2017; pp. 2174–2182.
33. Ren, S.; He, K.; Girshick, R.; Sun, J. Faster R-CNN: Towards Real-Time Object Detection with Region Proposal Networks. *Adv. Neural Inf. Processing Syst.* **2015**, *28*, 91–99. [CrossRef] [PubMed]
34. Russakovsky, O.; Deng, J.; Su, H.; Krause, J.; Satheesh, S.; Ma, S.; Huang, Z.; Karpathy, A.; Khosla, A.; Bernstein, M.; et al. ImageNet Large Scale Visual Recognition Challenge. *Int. J. Comput. Vis.* **2015**, *115*, 211–252. [CrossRef]
35. Wu, Y.; Lim, J.; Yang, M.-H. Online object tracking: A benchmark. In Proceedings of the IEEE Conference on Computer Vision and Pattern Recognition 2013, Portland, OR, USA, 23–28 June 2013; pp. 2411–2418.
36. Lim, S.C. Visual Object Tracking Algorithm Using Partial Fourier Siameses Network with Fully CNN-RPN. Ph.D. Thesis, Sunchon National University, Sunchon, Korea, 2022; pp. 1–132.

MDPI
St. Alban-Anlage 66
4052 Basel
Switzerland
www.mdpi.com

Energies Editorial Office
E-mail: energies@mdpi.com
www.mdpi.com/journal/energies

www.ingramcontent.com/pod-product-compliance
Lightning Source LLC
Chambersburg PA
CBHW061617210926
51850CB00009B/2354